Quantitative Methods in Construction Management and Design

Other engineering titles from Macmillan

J. G. A. Croll and A. C. Walker: *Elements of Structural Stability*
R. T. Fenner: *Computing for Engineers*
J. A. Fox: *An Introduction to Engineering Fluid Mechanics*
R. J. Salter: *Highway Traffic Analysis and Design*
J. D. Todd: *Structural Theory and Analysis*
E. M. Wilson: *Engineering Hydrology*, Second Edition

Quantitative Methods in Construction Management and Design

J. F. Woodward
B.Sc., M.I.C.E., F.I.Struct.E., M.B.I.M.

Director, Industrial Projects
University of Stirling

First published 1975 by
THE MACMILLAN PRESS LTD
London and Basingstoke
Associated companies in New York Dublin
Melbourne Johannesburg and Madras

SBN 333 17720 7 (hard cover)
333 18602 8 (paper cover)

Set in 10/12 Linotron Times and
printed in Great Britain by
J. W. ARROWSMITH LTD
Bristol

Contents

Foreword

by Professor Hugh B. Sutherland

Cormack Professor of Civil Engineering, University of Glasgow

It is often alleged that the construction industry lags behind other industries as regards the research that it fosters and the quality and methods of its management. To my mind these are unfair allegations—the construction industry is different from other industries. Most of the projects with which it is concerned are unique and cannot readily be tackled or managed by the direct application of production and other techniques that have been devised for mass production in the manufacturing industries. Many of these techniques, however, have been modified and applied to construction, mainly through the efforts of civil engineers such as the author of this book.

John Woodward is particularly well suited to write a book of this nature. After a distinguished undergraduate career he had extensive experience in the design research and construction sides of the industry. The later part of his time in industry was concerned with the development of construction management techniques and their application. His present appointment as Director of Industrial Projects at the University of Stirling has allowed him to maintain a foot in both the academic and industrial camps, and this blend of experience has resulted in a book that combines the best features of both approaches.

It is a book that is appropriate to the teachers of construction management in polytechnic and university courses and will also appeal and be of substantial assistance to practitioners in the industry.

It is with much pleasure that I have responded to the invitation from my friend and former student to write a foreword to what I consider to be an important and timely contribution to the literature of the construction industry.

May 1975   H.B.S.

Preface

The construction industry is based upon the skills of both technologist and entrepreneur. The former may be architect, engineer or other specialist contributing his particular skills to a complex interrelated industry. In his turn the entrepreneur, usually as contractor, provides the organisational ability to turn designs into reality. In the early days of the industry both functions were undertaken by the same man, then followed a period during which the tasks of design and construction became separated; recent years have seen a partial return to the co-ordination of all stages of a construction project. This latest development has been concurrent with the emergence of management science as a subject, based mostly upon applications in manufacturing production. A considerable body of knowledge and approach to problems has evolved under such names as work study, operational research, investment appraisal, decision theory, and so on; these have so far not been widely applied in the construction industry.

The purpose of this book is to introduce some of the methods of management science with particular emphasis on those methods that are of special relevance to the construction industry. The methods are all explained from basic principles, and no prior knowledge of them is required; emphasis is placed on illustrating their application by reference to practical examples, including many from the author's experience. In construction the most commonly used quantitative management technique is the critical-path method which has received a great deal of attention from other authors. For the sake of completeness it is included in this book, but only as one of a wide range of methods which are applicable to the industry.

It is hoped that the book will be of value both to students and to practising architects, engineers and construction managers. It has particular relevance to courses in building and construction management in both technical colleges and universities, as well as to those studying for corporate membership of the professional institutions. Since many of the applications described are new they may be unknown to those architects and engineers whose qualifications were obtained some time ago. Finally it should not be forgotten that this book also has something to offer to the many who have no formal qualification but are major contributors to the continued operation of the construction industry.

Projects are becoming ever more expensive, complex and interactive, and the rational approach offered by the methods of management science can

potentially provide benefits to all those who are prepared to make the effort
to tackle their problems in a methodical way.

The author wishes to acknowledge the inadvertent contributions made to
this book by his colleagues over the many years spent in the construction
industry, mostly with the Taylor Woodrow Group. Thanks are also due to
Professor S. L. Lipson, who provided the opportunity for the preparation of
some of the material while the author was Visiting Professor in the Department of Civil Engineering, University of British Columbia. Finally a debt of
gratitude is due to Mrs T. Gourlay for her untiring efforts in the preparation
of the manuscript.

Stirling,
October 1974   JOHN F. WOODWARD

Part One   Quantitative Methods and Concepts

Quantitative methods have been widely used in solving technological problems in the construction industry for a long time. It is relatively recently that the methods of management science have been applied to the industry, and the development of its application is described in chapter 1. In the last two decades the construction industry has made use of work study and network planning, but this has almost been the sum total of its use of management science. Chapters 2 and 3 review some of the more important aspects of the problems of construction planning, and the use of quantitative methods in planning. Parts two and three of this book go on to explore other areas of application, and for an appreciation of these applications it is necessary to understand the concepts of certainty, uncertainty, probability, variability and risk as explained in chapter 4.

1 Introduction to Quantitative Methods

Construction embraces building, civil engineering and plant erection. Building covers the whole range from single houses through housing schemes, offices, shops, schools and hospitals, to complex city-centre developments. Civil engineering is concerned with roads, railways, drainage, water supply, coastal protection, flood prevention, harbour work, power generation and mineral extraction. Building, civil engineering and plant erection come together in the construction of chemical plants, oil refineries and the whole range of other industrial undertakings including the energy-generating and distribution industries. Any individual product of the construction industry could be as small as a few hundred pounds in value in the case of small domestic structures, or as large as the multi-million-pound installations for power generation or oil-production platforms. In addition to covering a broad range of size of product unit it is a feature of the construction industry that it also covers a very wide range of skills—architect, engineer, surveyor and many different types of contractor and materials supplier. Most of these skills are organised in separate companies or units which means that in any one project there may be a large number of organisations involved. This fragmentation is a feature of the construction industry influencing the way in which it operates. Another specific feature which is typical of the construction industry is that most of its products are unique since with the exception of repetitive housing, most projects are of the one-off type. These characteristics of variability and uniqueness of product together with the fragmentation of the organisation of the industry make construction different from other industries. It should be noted that being different does not necessarily make working in the industry more difficult although many construction managers would claim that this indeed is so.

Much of the work in the construction industry is of a quantitative nature; the architect is much less preoccupied with aesthetics than is commonly thought and is more concerned with floor areas, building volumes, distances to fire exits, conformity with building standards, thermal conductivity and compliance with standards or norms laid down by building clients. The structural engineer is primarily concerned with the strength and stability of a building structure and undertakes detailed quantitative analysis of loadings, stresses, deflections and vibrations; in doing this he may well use complex analytical methods possibly with the aid of a computer to carry out lengthy calculations. The quantity surveyor, as his title implies, is responsible for the calculation of the quantities of materials and work to be undertaken in a construction project but he is also very much concerned with the costs of the work and with advising the client on the total financial position of the project; his work therefore borders on that of the accountant, but quantitative analysis is largely confined to calculations of an arithmetical nature. The contractor, among his other duties, will be involved in the calculation of quantities and costs, production planning and programming, as well as technical calculations associated with the design of temporary works and structures. The pattern in civil engineering is somewhat similar to building except that the designer in this case is usually a consulting engineer rather than an architect. The design of motorways, drainage systems, dams and earthworks, all require complex calculations and demand a high standard of mathematical ability on the part of the designer.

From these statements it is very obvious that a large part of the work in the construction industry is of a quantitative nature, and a large part of the teaching of skills for the construction industry involves computational methods and mathematical analysis. These are largely the *technological* skills of the industry and it is not the purpose of this book to review or repeat these. It is the intention here to explain and illustrate the applications of quantitative methods that have not so far been very widely used in the construction industry. Many of these methods come within the general subject-area of management science and more specifically they include work study, operational research, investment appraisal, business policy and technological economics. These subjects have been developed in industries other than construction, mostly in manufacturing or processing industries. Many of them have been concerned with production management and it is therefore largely in the management field that the methods have been applied and developed. Subsequently it has become apparent, however, that many of them can well be applied to design situations and therefore in later chapters of this book examples are given of both types of problem, namely design and management.

Much has been written[1] on the subject of the science and art of management, and it is not appropriate to discuss them at length here. There is a widely held view that management is an art that can be neither taught nor learnt, but is an ability that some people have while others have not. This belief holds that the skills and abilities of the 'haves' can be developed whereas the 'have-nots' will never become managers irrespective of how much training they are given. It is also part of this belief that the development of management skills is only possible in the hard school of experience in which the art can be developed. While there is some truth in the 'art' theory of management it is not the whole truth since it is now becoming accepted in many industries that the science of management has a valuable part to play. It is however only a part, and the best results will be obtained when art and science act in a complementary way. It may at times be unwise to rely on hunch or intuition when making management decisions but it is equally unwise to go to the other extreme and rely on the methods of management science entirely and ignore the intuition of the experienced manager. It cannot be stated too often that management science can provide useful tools of management, providing managers with assistance in the complex problem of managing a business or project. It is not claimed that management science can supplant management art and anyone who thinks that it can is in danger of making great mistakes. The methods of management science can simply offer to managers the means of improving their decisions and actions.

Operational Research

Within the construction industry applications of operational research have largely been confined to network planning methods. The fundamentals of critical-path method are covered in chapter 2, at the end of which is a list of references that extend the subject considerably further. Network planning has been included because it is thought that no book on quantitative methods in the construction industry would be complete without it, but it is essentially the purpose of this book to show that there are very many other problem areas where the methods of management science can usefully be applied. For this reason great emphasis is given to problems of uncertainty and probability and the discussion of the techniques of linear programming, transportation, cost models, investment appraisal, decision theory, dynamic programming and competitive bidding strategy. This emphasis is not intended to diminish the importance of network planning but simply to illustrate that it is only one of many techniques and it is the one that has been given the greatest prominence in the literature to date.

Operational research, hereafter abbreviated to OR, had its origins in Britain during the Second World War. Its origins, development and coverage are well described elsewhere,[3,4] and it is sufficient here to give a brief description of the concept. OR was originally undertaken by an interdisciplinary team, mostly scientists, set up to examine some specific operational problems of warfare.

The classic example is that of the effectiveness of depth-charges against submarines. Defence officers were becoming increasingly alarmed at the low success-rate of anti-submarine action in spite of a high success-rate of submarine detection. It was known from technical tests that the maximum effective radius of a depth-charge occurred when it was detonated at a depth of about 30 metres, and therefore all depth-charges were set to detonate at that depth. The essential point that was missed however, was that the majority of submarines did not cruise at a depth of 30 m but were generally nearer to the surface. Once this fact was recognised it was possible to assess the chance of success for each detonation depth, taking account both of the effective radius of the depth charge *and* the probable cruising depth of the submarine. This assessment was first of all made by calculation using a *model* in order to find an *optimal* solution to the problem. The solution worked out in this case showed that depth-charges should be detonated much closer to the surface and it was decided to undertake a series of operational tests (incidentally, after much argument among the people concerned). A considerable improvement in the success rate was achieved by reducing the detonation depth, and the principle was adopted for general use.

This description, although very brief, does bring out some of the essential features of OR as follows.

(1) The existence of a problem area in which there is an *objective* which seeks to find an *optimal* solution. In this case the objective was to maximise the effect of depth-charges against submarines, but in most industrial situations the objective is to maximise profits or minimise costs or minimise time, and so on.

(2) The use of a *model* which represents the system under study. In the submarine case the model consisted of a series of relationships between depth and frequency of submarines, detonation radius, or chance of success. In industrial situations the model may consist of a bar-chart programme, a network plan, a sales chart, or a productivity formula.

While these two features characterise OR the other essential steps in the study that are common to all management-science techniques are as follows.

(3) Definition of the problem area. It is necessary to be firm about the definition of the problem area since it is very easy for the problem to grow and grow to such an extent that a solution becomes unobtainable. Again referring to the submarine problem, it was clearly defined that the purpose of the study was to improve the success rate of depth-charges against submarines. A slightly wider statement of the problem would have been to seek a better method of destroying submarines, but this would have opened up possibilities of using other methods and would have greatly complicated the problem. A still wider interpretation of the problem would have been to improve the survival rate of shipping convoys reaching Britain, and this could possibly have been achieved by other means such as the use of different cargo ships or changing their routes. An even wider interpretation of the problem would be to improve the arrival rate of food and other essential goods and materials into Britain during the war. Each stage of enlargement of the problem may well justify detailed study, but the wider it is the longer it will take to solve and to implement the results. It was obviously valid to carry out the depth-charge exercise and implement the results quickly since these had an immediate effect and did not prohibit subsequent study of the wider problem areas.

In rather the same way a contractor may wish to optimise his concrete-handling methods on a large site, and if this is all he wants to do, care must be taken to ensure that the problem is carefully defined as such. If no careful definition of the problem is made it could easily embrace the planning of all operations on the site, and may even extend to the planning of operations on other sites. Again wider consideration may be justified in some cases and indeed if the problem area is too closely defined and isolated there is the possible danger of sub-optimisation. Care must be taken to ensure that implementation of the optimal answer for concrete handling does not interfere so much with other operations on the site that the overall total situation represents a deterioration.

(4) Collection of information. Before tackling any problem it is obviously important to collect data that is adequate both in terms of quantity and quality. It is a truism to state that no solution can have an accuracy greater than the accuracy of the data put into the problem, but there is a danger that the use of an OR method will itself imply an accurate answer.

(5) Testing the solution. It is usual to test the solution on the model which represents the system but it is just as important, if not more so, to test the solution under actual working conditions. This will guard against the situation where the model does not truly represent the system, and factors which were not thought to be important turn out to have a significant effect.

(6) Implementation. When all the other steps of the study have been completed it is necessary to implement the solution and thereafter to make sure that it continues to operate as expected. This may mean a formal review, since it is possible that changes in the system may occur with time, leading to a change in the optimal solution.

Applications of OR and Management Science

Many applications of OR are discussed in the later chapters of this book, but there are many techniques and methods that are of interest to a whole range of industries, including construction. Many of these have been omitted from this book either because they are of highly specialised application, or because their application is so general that it is a very simple matter to adapt them from other contexts. They are simply listed here to show the wide range of problems in which OR can play a very useful part.

(1) Statistical analysis is widely used in quality control. The quality of building materials is mostly controlled by manufacturers and suppliers, but one area in which statistical quality control is commonly used by construction engineers is in the use of concrete on site.[2] Another area in which statistics are less commonly used is in forecasting.

(2) Stock control. OR methods are commonly used to optimise the purchase, handling and control of stocks, and there are many instances in construction where these methods are applicable.

(3) Flows in networks. Electrical, hydraulic and traffic engineers may be familiar with the analysis of flows in networks, and many of their methods may by analogy be applied to problems of production and transport flow.

(4) Queuing theory. This complex mathematical theory was originally developed in the design of telephone systems, but has subsequently been applied to aircraft arrival at airports, passenger-handling systems, service points in post offices and similar problems.

(5) Technological forecasting is concerned with rates of adaptation of new technology. It is probably therefore of more interest in the design of buildings and community services rather than in their detailed construction.

(6) Maintenance. The use of quantitative methods to assess the alternative systems of planned maintenance and breakdown maintenance, and their effect on plant reliability and availability. The methods used in manufacturing industry are readily applicable to construction plant and equipment.

Models

It has already been stated that a common feature of management-science techniques is the use of a model in tackling a problem. A model is used first as a means of helping to understand how a system operates, and it can give a much better description of a system than could possibly be done with words. Models are also used because it is possible to carry out experiments upon them without the costs and consequences of experiments on full-scale systems. For example in construction planning we use a simple bar-chart programme as a model to represent the time scale of the project. This will show separately the time required for foundations structure, finishes and services of a building, broken down into a number of component parts. If consideration is being given to two or more alternative methods of construction this bar-chart model can give an indication of what effect each of the methods will have on the overall completion date for the project.

A bar chart is in fact a very simple model and perhaps an inaccurate one, and it is likely that a much better understanding of a construction project will be given by a network diagram as is described in chapter 2. There are three broad categories of models which may be used, namely descriptive, quantitative and interactive.

(1) Descriptive models include drawings, bill of items, and physical visual models. Even though a drawing may be produced to exact scale and may carry dimensions, it is still only a descriptive model since it simply determines and describes the size of the building.

(2) Quantitative models in their simplest form consist of graphs or mathematical formulae. They usually involve at least one mathematical variable and can be used to find values for that variable under particular

circumstances. For example the following simple formula would be a model representing the bonus payable to operatives for early completion of a particular operation.

$$B = 0 \cdot 30(t_t - t_a)r$$

where

B = bonus payment
t_t = target time for the operation in hours
t_a = actual time taken for the operation in hours
r = basic hourly wage rate

This form of incentive formula is thought by some people not to offer sufficient incentive at the lower rates of production; if operatives feel that there is no reasonable possibility of improving significantly upon the target time then there is no incentive for them to come anywhere near the target time. Accordingly in many situations a rather more complex relationship between output and bonus is evolved and this is most easily presented in the form of a graph, that is, a simple form of model as shown in figure 1.1 by the full line. For comparison the dashed line shows the relationship corresponding to the previously given formula. Both lines offer similar bonus rates over the normal working range of output from 0·5 to 1·0 square metres per man-hour, but the curved full line offers some incentive in the range 0 to 0·5 square metres per man-hour, but at the same time removes the temptation to attempt very high output rates with the consequent possibility of poor quality of work. It is not appropriate to discuss this point further here, but it

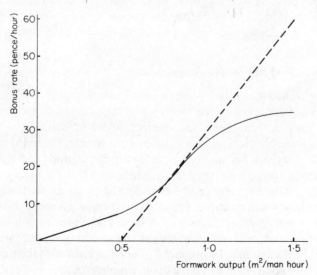

Figure 1.1   Bonus rate as a function of output rate

can be seen that the comparison of various incentive schemes on a graphical model such as this provides a ready answer to questions like 'What happens if . . .?'.

More complex quantitative models can represent the interaction of several variables, for example in the case of concrete-mix design. Here the interaction of aggregate/cement ratio, water/cement ratio, aggregate grading, workability, and ultimate strength can be represented not on a single graph but in a series of graphs. Still more complex interactions can be described and evaluated using the models and methods of linear programming described in chapter 5.

(3) Interactive or dynamic models are usually of a quantitative form but have the additional feature of representing not a static problem situation but a changing one where the original model can be modified at various stages by input data. A critical-path network is an example of this where it represents all the individual tasks that fit together to make up a construction project; it is possible to feed into this model the effect of one of the early tasks running late, and the modified model will represent the new situation. In some industries very complex interactive models have been constructed—one of the most significant being in a major oil company which has a mathematical model that represents the whole of the company's operations, including bulk crude transport, refining, sales and distribution. This model can answer questions of the type 'A tanker of 100 000 tonnes capacity carrying crude oil of type X has just left Saudi Arabia, four days behind schedule and is expecting to run into bad weather; to which refinery should the tanker be sent and what should the allocation of the crude be to the various possible refining processes?'. This is a rather more ambitious model than would normally be found in the construction industry but it does indicate the possibility and scope of the methods of mathematical modelling.

Quantitative and Subjective Assessment

Engineers traditionally like a deterministic solution of problems, and they are happy when they can carry out a set of calculations to produce a single answer. They are confident and competent when handling data that is readily measurable such as linear dimension, weight, strength, speed, frequency. It is perhaps because of this deterministic attitude that they are less confident when faced with problems in which they have to make a subjective judgement. A site engineer can be confident of giving an accurate answer if he is asked how much concrete is required to complete a particular structure. He is much less confident, however, if he is asked how long it will take to carry out the work, and will probably give an answer of 'about three weeks'. If pressed further he may be prepared to say that it should take three weeks, there is a very good chance that it will be completed by four weeks, but there is a slight possibility that it might take five or more weeks. Few construction managers will be even as explicit as this, and very few indeed will be prepared to give a more precise and quantitative estimate of the time required for the completion of a project. Experience over a large number of contracts should enable a competent manager to state that there is a 50 per cent chance of completing the work in three weeks, a 75 per cent chance that it will be completed in four weeks and a 90 per cent chance that it will be completed in five. It should even be possible to state that there is a 2 per cent chance that the work could take as long as ten weeks. It is one of the purposes of this book to introduce and illustrate the notion that factors that are normally regarded as being purely subjective can be given a quantitative value as in this example.

One situation where this can obviously be done is in relation to bad weather on a construction site. It often appears to come as a surprise to some construction managers that they suffer a spell of wet weather which causes a delay to their progress. Very good statistical information is kept for all parts of the country permitting an accurate estimate to be made of the chances of bad weather. For example records are available to show that for a particular site the chances of work being 'rained off' are shown in table 1.1.

This shows not only that in winter there are on average three days per month lost, but that more than seven days could be lost in a month; in the summer period the average number of days per month lost is about $1\frac{1}{2}$, and is not likely to exceed five days. This gives a much more specific picture of probable impact of weather upon a construction site, it is based on good statistical information, and it should be used as a basis of planning.

Table 1.1   Percentage chance of a number of days 'rained off' in a month

Month	Days lost							
	0	1	2	3	4	5	6	7
Jan	2	10	22	34	20	6	4	2
Feb	2	9	22	35	20	6	4	2
Mar	2	9	23	37	18	5	4	2
Apr	5	11	24	35	15	5	3	2
May	7	16	25	35	9	4	3	1
June	10	22	28	25	9	4	1	1
July	14	31	28	16	8	3	0	0
Aug	15	30	28	16	8	3	0	0
Sep	16	30	30	14	7	3	0	0
Oct	12	22	26	28	6	4	1	1
Nov	8	21	26	30	7	5	2	1
Dec	5	16	24	36	9	5	3	2

Confidence Limits

In many calculations it is not possible to give a precise answer, either because the method of calculation has not been precise or because the input data has been approximate. It is useful to be able to quote the degree of confidence that can be placed on the results of such calculations and again this can be something rather more quantitative than simply saying that an answer is either approximate or accurate.

Consider the simple case of the calculation of the volume of a cube of concrete, the face of which is not perfectly regular. It is assumed that the irregularities are such that the dimension of the cube can be measured to within ± 10 mm for a cube the nominal size of which is 1 m. This amounts to saying that the *confidence limit* of the length of the side is ± 1 per cent. The nominal volume of the cube is $1 \cdot 0 \times 1 \cdot 0 \times 1 \cdot 0 = 1 \cdot 0 \, m^3$, but the *smallest possible* volume is $0 \cdot 99 \times 0 \cdot 99 \times 0 \cdot 99 = 0 \cdot 9703 \, m^3$ which is 3 per cent lower than the nominal volume. The largest possible volume is similarly 3 per cent greater than the nominal volume, or in other words the confidence limits of the volume of the cube are ± 3 per cent.

Where calculations are based on approximate input data it is quite likely that the confidence limits of the answer are very wide indeed, and for this reason it is usual not to take the whole of the range over which the value may lie but only part of it. Based on statistical principles it is common practice to talk about 90 per cent confidence limits, meaning that in 90 per cent of the cases to which it applies the solution will lie within the limits quoted, but there will be 10 per cent of cases where the answer lies outside these limits. In practice this permits a much narrower range of limits to be quoted while still giving an indication that less accurate results would apply in a small number of cases.

It is now fairly common practice to make use of confidence limits in the assessment of proposed investments where management is very keen to know not only what the possible pay-off from the investment could be but over what range of values this pay-off could possibly lie. It would be appropriate to apply similar confidence limits to the preparation of estimates for construction work but this approach has not yet been adopted in very many cases. It would for example appear to be sound practice never to submit a tender for a contract that had a value below the upper confidence limit of cost of completion of the work; this would ensure that no contract was ever undertaken at a loss but of course at the same time it might mean that the company was consistently bidding too high to obtain any contracts, especially if the confidence limits on cost estimate were wide.

Management

One of the basic principles of management is particularly applicable where quantitative methods are being applied. Whatever the industry or whatever the situation one of the fundamental tasks of management personnel is to exert control over the work. This function can be schematically represented in figure 1.2 which shows that the starting point of any operation is to *plan* what is to be done. As work proceeds care must be taken to *measure* what is

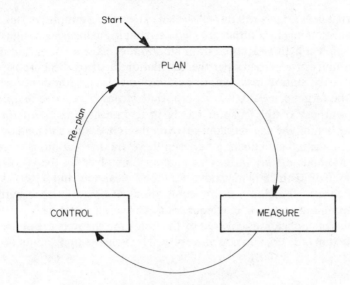

Figure 1.2   Management-control cycle

being achieved, and thirdly when the work done is compared with the plan it is necessary to exert *control*. This is not the end, however, since the process has to be repeated and does indeed have to be recycled.

In terms of construction work this can be applied in many fields. In relation to progress on a construction site a plan or programme is prepared at the start of the contract and then as work proceeds a measurement is made to check the progress of work compared with the plan. If there are discrepancies between the plan and actual site progress it is necessary to exert control by applying additional resources, removing obstacles or otherwise ensuring that work is accelerated and brought back into line with the plan. If it is not possible to bring the work schedule back into line with the original plan it may be necessary to revise the plan, which then serves as the basis of comparison against further progress.

Quality control of concrete can be regarded as another example where a certain quality of concrete is specified in the contract, a plan or design of the concrete mix is undertaken and measurements are made at intervals by testing sample cubes. If these cubes do not show that the specified concrete strength is being maintained then it may be necessary to revise the design of the mix in order to achieve the required objective.

Cost control on projects is in exactly the same position where a target cost for a particular operation is set, the actual cost of carrying out the work is carefully measured or costed, and finally control has to be exerted in order to bring costs into line with the target possibly by changing working methods or work supervision. If it proves to be impossible to meet the cost target it is likely that new targets will have to be set. There are examples of this sort of approach not only in contract management but in engineering design, where for example a hydro-electric power station may have a target output of X million units of electricity per week. Measurement of the output performance of the station may lead to modification of its methods of operation and if the target proves to be inappropriate it may have to be revised.

The purpose of this chapter has been to introduce some of the basic thinking behind the use of quantitative methods in design and management, and to indicate the distinction between these and the traditional computational methods of the various technologies involved in the construction industry. The distinction may not always be a clear one, and this book by no means covers all of the possible applications of operational research and other techniques to the management of construction, but it is hoped that the examples of applications covered by the following chapters *are* of use in the construction industry and not 'all very well for other people, but of no use to us'.

References

1. P. F. Drucker, *The Practice of Management*, Heineman, London, 1955.
2. R. Oxley and J. Poskitt, *Management Techniques applied to the Construction Industry*, Crosby Lockwood, London, 1968.
3. P. Rivett, *Concepts of Operational Research*, Watts, London, 1968.
4. R. L. Ackoff and M. W. Sasieni, *Fundamentals of Operations Research*, Wiley, New York, 1968.

2   Network Planning Methods

The construction industry has been slow to take up a wide range of operational research techniques, but one area that has become widely developed is the use and application of network planning methods. There is no generally accepted name for these methods but they are variously called network planning, CPM (critical path method), PERT (programming evaluation and revue technique) and other similar names. There have been many useful books published on this subject, some of which are referred to at the end of this chapter,[1,2,3,4] and it may be superfluous to go over this material yet again in this book. The use of the methods has however become very widespread in the industry and it is therefore felt that no book on the use of quantitative methods in the construction industry would be complete without a chapter on network planning methods. This chapter sets out to explain the fundamentals of network planning methods and to discuss the advantages, disadvantages and practical applications of the methods. It is not possible within a single chapter to cover the whole ground and therefore the more sophisticated and advanced aspects of network planning methods have been omitted, but the reader is directed towards other books on the subject.[5,6]

While it is true that network methods have been widely adopted throughout the construction industry, it is also true that there still exists much suspicion of the method, and there is great scope for its better use, and for extending it into companies and organisations who are currently not convinced of its value. This suspicion of network methods has derived in many cases from experience of attempted applications of the methods which have led into difficulty if not into disaster. The natural reaction to failure in an attempt to use a network is to reject the method; but what is often forgotten is that it may not be the method that is at fault but the way in which it has been applied.

To take a very simple analogy, a construction manager who finds that one of the buildings under his control has been constructed one metre longer than indicated on the plans, will not draw the conclusion that tape-measures do not work and should be discarded. The same construction manager who finds that his project is running one month behind schedule should therefore not blame the planning methods being used; in the first case it is certain that the tape-measure is being wrongly used, in the second case it may be that the planning method is being wrongly used, or alternatively that despite the planning method it has proved impossible to complete the works according to a schedule that was unrealistic in the first place. It is becoming evident, through more than fifteen years' experience of the use of network planning methods, that despite many set-backs the method is a powerful one and offers many advantages to project managers and others concerned with the control of construction and similar works.

Some of the advantages and disadvantages of the methods are discussed later in this chapter but there is one overriding general advantage which singles out network methods from other construction planning systems. This is that network methods offer the only true way of representing the logical dependency of sequential parts of a construction project. Whether or not a network is drawn, the separate elements of a project fit into some logical sequence, for example in the construction of a simple brick wall it is necessary first of all to excavate for the foundation, then concrete the foundation, then build the wall—all in that sequence. A bar chart could represent this programme but would not indicate that each operation *must* be complete before the following operation can commence, whereas a network does precisely this. This may not at first appear to be very important but in a large project where there are many interacting operations to be completed it is difficult to envisage how they will interact without the fairly precise methods offered by networks. In the construction industry we are often given a required completion date and then seek to compress the project to fit within the time available. It is very easy to delude ourselves by simply compressing or overlapping the bars on a bar chart, but it is much less easy in practice to compress or overlap operations on a construction site. The use of network methods will point out where it is possible and where it is not possible to overlap operations, and where the reduction in time required by individual operations offers the greatest time saving to the overall project. These points will become clear later in the chapter where they are further discussed.

Arrow Diagrams

Many different versions of network planning methods exist, the most common being CPM and PERT, which have certain distinctive features that are irrelevant to the fundamental approach to network planning. Many other versions of the method exist, most of them being concerned with either particular applications or with the use of computer methods. A common fallacy is that network planning requires the use of a computer, but this is not true. Many large applications of network planning do in fact use computers, but for every large network put on a computer there will exist at least ten smaller networks which are entirely operated by hand. Within the whole range of network planning methods there are however two fundamentally different approaches to the preparation of a network and it is important that these two methods are kept distinct; an analogy here would be that it is perfectly reasonable for vehicles to drive either on the left-hand side of the road as in the United Kingdom or on the right-hand side of the road as in Europe; each system works well on its own, but a mixture of the two would produce extreme difficulty. In the same way we may use either the so-called arrow-diagram method or the so-called precedence-diagram method. In order to avoid confusion in this book only the former will be discussed, and anyone seeking information on precedence-diagram methods should refer to other suitable texts.

In order to understand the construction of a network we need to define a number of terms. The following terms are standard throughout all versions of arrow-diagram planning but the definitions are the author's own interpretation of these terms.

Activities and Events

A *job* or *activity* is the individual operation which for the purpose of planning can be considered as a unit, and does not need to be subdivided. A *project* is the overall undertaking which is to be represented by the entire network.

An *event* is a point in time at which preceding activities are complete and succeeding activities are not yet started. It is represented in a network by the junction between arrows, and in network terminology is called a *node*.

In tabular form

	On site		On paper
a *job* or *activity*	is represented by	an *arrow*
a *project*	is represented by	a *network*
an *event*	is represented by	a *node*

Each job is represented in the diagram by an arrow, the tail indicating the point at which that job *can* start and the head the point by which it *must be* finished. Since the network is intended to represent first the sequential relationship and not the length of time taken, the length of the arrow and its orientation are not significant, and any of the forms shown in figure 2.1 are equally valid. It is, however, convenient to draw arrows from left to right. In order to place an arrow in a network it is necessary to ask the following three questions.

(1) What other jobs must be complete before this one can start?
(2) What other jobs can be done at the same time?
(3) What other jobs cannot start until this one is finished?

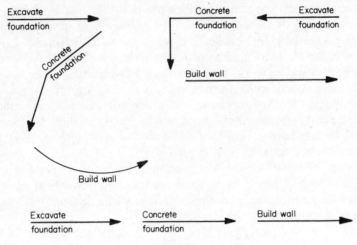

Figure 2.1   Various forms of simple network

Construction of a Network

Consider the simple example of erection of a portal frame (figure 2.2), in which the operations are limited to the following

> Clear site
> Prepare base A
> Prepare base B
> Erect column A
> Erect column B
> Erect beam
> Plumb and line structure
> Grout up bases

Consider first the job of clearing the site. We ask the three questions listed above and find that no other job need be complete first, no other job can be done at the same time, and all other jobs cannot start until this is complete. It is then clearly the first arrow we draw and must precede all others.

We then look to see which job can now start, and see that the preparation of base A can commence. Again the three questions are asked, the answers being

> (1) Clearing the site must be complete
> (2) Preparation of base B may proceed concurrently
> (3) Erection of column A cannot start until this job is finished

This enables us to draw in the arrow for the preparation of base A, starting at the head of the arrow for the site clearance. Another arrow for preparation of base B can start from the same point.

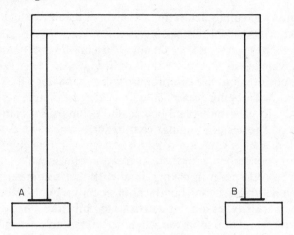

Figure 2.2   Portal-frame project

The same rigorous questioning of all activities enables us to complete the diagram as shown in figure 2.3. Note that the erection of column A depends only on base A and not on base B. At the same time the erection of the beam can only follow the erection of both columns. Remember that the network should show what is reasonably physically possible, without taking special measures; for example it would be possible to erect the beam without one of the columns by using temporary supports, but this would entail special measures and should not be shown unless these measures are required for other reasons.

Level of Detail

The principle outlined above is extremely simple, but in its simplicity there is a series of problems. The first problem is that in writing down the activities we have established a level of subdivision of the project into its elemental

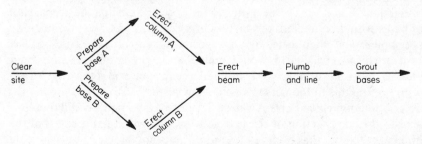

Figure 2.3   Portal-frame network

parts. It is often difficult to determine the level of detail to which a network should be taken; if the level of detail is very fine we have a large network with very many activities, whereas if we do not go to a fine level of detail the result is a network which is so general that it is of little value. There is as always a compromise position, but this compromise will depend upon the purpose for which the network is being drawn. There would, for example, be little point in going to the level of individual beams and columns associated with the construction of a series of buildings comprising a new town development. The network in such a situation would be enormous. It is equally obvious that if we were concerned with the construction of a school in a new town there would be little point in having a network that comprised only three activities—namely plan school, build school, occupy school—although these might well be perfectly reasonable activities to build into a network used to control the overall planning of the whole town. A decision on the appropriate level of detail in a particular case can only really come from experience from the use of networks. As a very rough guide it might be said that if the time required to complete any individual operation is less than one-thousandth part of the project time, then the level of detail is too great; conversely if the time to complete one operation is more than one-tenth of the project time then the level of detail is too shallow.

Another problem arising at this stage of drawing the network is that there is no one correct way of drawing a network, and different planners will prepare different networks representing the same project. It could be argued for example that in any real project there would only be one gang of men available to prepare the two bases in the problem shown in figure 2.3 and it might be tempting to indicate that the preparation of base B should follow on after the preparation of base A. This would, however, indicate that it is not possible to prepare base B until base A is completed, which of course is not true. The sequencing of activities in this way is really a matter of convenience of operation rather than of true logical dependency. At the early stages of the preparation of a network it is important that logical dependency be the primary consideration, the sequencing of activities to suit resource limitations can be done at a later stage, that is, at the stage when a plan of action is being prepared. There could be a converse argument which says that it is likely on a project such as the portal frame that the preparation of bases would be undertaken first by bringing on a hired excavator to dig two holes, and then bringing in a ready-mixed-concrete truck full of concrete for two bases. This would imply that the preparation of base A and base B must be done as a combined operation, and the only way of representing this in the network is to show a single arrow with the activity description of *prepare bases A and B*. These two arguments could therefore lead to the construction of the two networks shown in figure 2.4, but it is important to note that each of these represents not a logical analysis of the sequential operations necessary for the project, but rather a statement of the

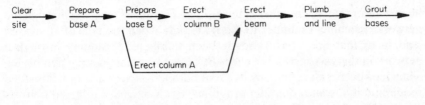

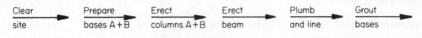

Figure 2.4   Alternative networks for portal frame

intended method of construction. This leads us to an important point that has been the cause of the failure of networks in many situations and is as follows.

It is important at the outset to prepare a network that represents the true logical dependency of activities one on the other, *without regard for the efficient use of resources or the desired method of construction*. This then gives a statement of what is technically feasible and this may then be used to calculate the time required for the completion of the project and will in turn lead to the preparation of an intended plan of action. It is important that these stages be kept separate.

While this example is extremely simple it serves to illustrate three main points

(1) The strict logic of one activity following on after another is completed.

(2) An activity is dependent only upon the jobs that *must* precede it, not also on jobs which *may* precede it.

(3) It is important to have an understanding of the project otherwise the wrong logic may be used.

Events

Each point of intersection in the network is called a *node* or *event*. The node is the point on the network as drawn, and the event is the corresponding point in the project, for example, in figure 2.3 the point at which both columns have been erected, but not the beam.

For reference purposes it is useful to number all events, and it is normal practice to start at the beginning and number all events in ascending order. While many computer programs can now handle completely random event numbering it is good practice to ensure that each activity has a number at its head greater than the number at its tail. This is ensured by going through the network assigning a number to the node only when the tails of all arrows leading to that node have already been numbered. Running through a network in this way serves as a check, it helps to have sequential numbering when looking for an activity by its event numbers, and in addition it is still a requirement of some computer programs. Each activity is referred to by its start and finish event numbers, for example, 2–3, 3–4, etc.

Dummy Activities

A *dummy* is shown in a network by a dotted arrow and is a device to maintain correct numbering and logic in a diagram. To illustrate this consider first the following very simple logic diagram

A is the first job
B and C follow A
D cannot start until B and C are complete

This would be represented and numbered as in figure 2.5. In this network jobs B and C both start at event 2 and both finish at event 3, and hence are both numbered as jobs 2–3. This could lead to confusion, and in order to achieve unique numbering a *dummy activity* is introduced as in figure 2.6.

This in no way changes the logical sequence of events, but now job B is 2–3 and job C is 2–4. A dummy is exactly the same as a real activity as far as the logic of a network is concerned, and it should always be treated as such, the only difference, as is described later, is that it requires zero time for its completion.

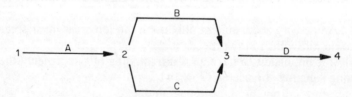

Figure 2.5   Duplication of activity reference

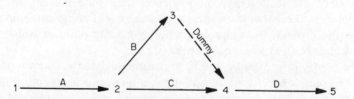

Figure 2.6   Introduction of dummy to give unique activity reference

There is a second and more important use of a dummy activity, where it is needed to clarify particular situations, for example, in the following logic

A and B are carried out at the same time
C cannot start until A and B are complete
D cannot start until B is complete

At first sight this can be shown as in figure 2.7. However, on inspection this also stipulates that D cannot start until A is complete, but this was not an original condition. By the introduction of a dummy we can correct this in figure 2.8, showing that while activity C does depend upon the completion of both A and B, activity D may proceed as soon as B is completed.

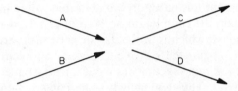

Figure 2.7   Inclusion of superfluous dependency

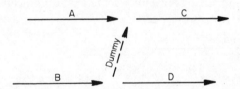

Figure 2.8   Inclusion of dummy to remove superfluous dependency

It is fairly easy to understand this use of dummy activities, but many people when drawing a network for the first time fall into the trap of building in false dependencies. A typical situation is shown in figure 2.9 which represents part of a large network where a number of activities come together representing the stage of the building project where the building is

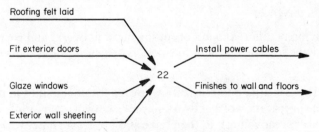

Figure 2.9   Part of a building network, illustrating the cluster problem

water-tight, the roof and walls being complete and the exterior doors and window glass installed. It is usually considered necessary that a building should be water-tight before the installation of power cables or work on wall and floor finishes can commence, and these are therefore shown on figure 2.9 as activities leading from the event representing the state of the building being water-tight. On a network this type of event, where many activities lead into it and many activities lead out of it, can informally be called a *cluster* event.

The representation shown in figure 2.9 is perfectly correct but it may subsequently be necessary to include an additional activity, for example the building of an internal partition wall which has to be attached to the wall sheeting. It is tempting at first sight simply to start an activity at event 22, in figure 2.9, to represent this partition wall, but that would indicate that it is also dependent upon the completion of the roofing felt, the exterior doors and the window glass, which is probably not necessarily true. The correct procedure, shown in figure 2.10, is to introduce a new event, number 21, at the completion of the wall sheeting but separating it from event number 22 by a dummy activity. The new activity—the construction of a partition wall—which does depend upon the external wall sheeting can then start at event 21, thereby showing that it does not depend upon the other activities leading to event 22.

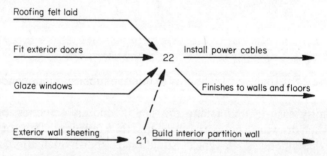

Figure 2.10   Clarification of the cluster problem

Laying out a Network

There are many different ways of setting out a network, and each planner will evolve the method that suits him best, but the following points may be of use.

(1) There is little point in writing down a list of activities in note form prior to the preparation of a network since the network can form its own list. There is also no advantage in the preparation of a rough network which is then subsequently re-drawn, and there may in fact be a danger since in the re-drawing or tracing process it is possible that errors of logic can easily

creep in without being noticed. After a little practice it should be possible to produce a reasonable working network at the first attempt.

(2) It is important not to draw the network at too small a scale since again this is a possible source of error. Much fun is poked at networks which occupy long rolls of paper, sufficient, we are told, to complete the papering of the walls of the site office, but planners must not be put off by such jibes. It is not difficult to handle a network on a fairly long roll of paper since it is likely that at any particular time only part of the network is required for examination and the earlier and later parts of the network can be rolled up.

(3) It is useful to show each arrow with a horizontal portion somewhere along its length, the horizontal portion being used for the activity description that should always be included in every network. This lay-out with horizontal lines is not simply a matter of neatness, but permits the network to be separated into a series of horizontal zones representing either separate sections of the project or perhaps different types of work, for example, design, structural, services, and finishes.

(4) Remember that the original network should be drawn on the basis of unlimited resources and should not be adapted to take account of preferred methods of construction.

Scheduling, Putting the Network to a Time Scale

So far only the logic of a network has been discussed and no mention has been made of the time required to complete the project. It is usually best to complete the logic network first and then to go back and assess the time required for each activity; this is called the activity duration which for the purposes of this book is being treated very simply. Some networking methods recognise that the time required for a particular activity cannot be accurately predetermined and may vary from the estimate made. The PERT system simply takes three time estimates—namely an optimistic, a pessimistic and a most likely—and combines these in the form of a weighted mean. This simple mean takes very little account of the variability of activity duration but the more rigorous combination of a series of probabilistic time estimates becomes mathematically rather complex. In this chapter therefore the single time estimate for each activity duration is used. It is usual to select a time unit which permits the use of whole numbers for activity durations and is a consistent unit; for example, rather than say that an activity takes one and a half weeks or one week and three days, we would translate the duration into the number of *working days* required; it is not therefore necessarily calculated on the basis of seven days per week. This can also lead to difficulties since we need to translate certain activities specifically into working days; for example, if on a particular contract we are required to give concrete twenty eight days' curing and we are only working a five-day week, the curing activity should be shown to have a duration of $4 \times 5 = 20$ working days.

Calculation of Project Duration

The method of calculation of the time required for the completion of the project is as outlined below and given in figure 2.11. There are many different forms of notation in use with network planning methods but the notation given here is one of the most common. Each activity duration is indicated in brackets below the line representing that activity. It is now necessary to define two terms that help in the calculation of project times, and these are as follows.

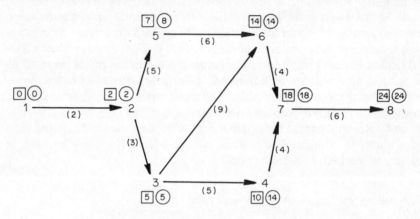

Figure 2.11   Calculation of earliest and latest event times

The *earliest event time* for each event is the earliest time at which that particular event can occur, that is, the earliest time by which all activities leading to that event can be completed. In the notation used here, the earliest event time (EET) is indicated in a rectangular box at each event.

The *latest event time* is the latest time at which each event can occur without causing a delay to the project, that is the latest time by which all jobs leading to that event must be complete. The latest event time (LET) is indicated in this notation by placing a number in a circle at each event. The event times are calculated as follows with reference to figure 2.11.

The project starts at time zero and hence we put this as the earliest event time for event number 1. Activity 1–2 has a duration of two days and therefore clearly the earliest time of event 2 is $\boxed{2}$. Proceeding through the events in ascending event number order we can now calculate that the EET of event 3 is $\boxed{2}$+(3)=$\boxed{5}$, and the EET of event number 4 is $\boxed{5}$+(5)= 10. When we come to calculate the earliest event time for event 6 however, we have a problem since there are two alternative ways of calculating it, namely $\boxed{7}$+(6)=$\boxed{13}$, or $\boxed{5}$+(9)=$\boxed{14}$. While the sequence 1–2–5–6 gives us the quickest route to event 6 we must remember that in order for event 6 to occur, all activities leading to it must be complete, hence we must choose the *longer* of the two alternatives to determine the earliest event time at event 6 namely $\boxed{14}$. Similarly at event 7 we have two alternatives and take the longest, that is $\boxed{18}$ and the earliest event time for event 8 is $\boxed{24}$. This figure of $\boxed{24}$ gives us the earliest date by which the overall project can be complete.

The calculation of latest event time is the reverse procedure for the calculation of earliest event time. Starting with the final event, number 8, the LET must obviously be $\textcircled{24}$, and if we are to be at event 8 by time $\textcircled{24}$ we must have a LET at event 7 of six days earlier, namely $\textcircled{18}$. We then proceed backwards through the network in descending event number order to calculate the LET at each event. We again have a choice when we come to event number 3 where the LET could be either $\textcircled{14} - (9) = \textcircled{5}$ or $\textcircled{14} - (5) = \textcircled{9}$. In this case we must remember that event number 3 must occur at such a time that there is sufficient time for all subsequent activities to be completed, and we must therefore take not the shortest route from 3 to 8 but the *longest* route, namely 3–6–7–8, giving us a LET at event 3 of $\textcircled{5}$. The full calculations for both EET and LET are set out in table 2.1.

Table 2.1   Calculation of EET and LET

Event number	Earliest event time	
1	0	
2	2	
3	$2 + 3 = 5$	
4	$5 + 5 = 10$	
5	$2 + 5 = 7$	
6	either $7 + 6 = 13$	
	or $5 + 9 = 14$	EET for event 6
7	either $14 + 4 = 18$	EET for event 7
	or $10 + 4 = 14$	
8	$18 + 6 = 24$	

Event number	Latest event time	
7	$24 - 6 = 18$	
6	$18 - 4 = 14$	
5	$14 - 6 = 8$	
4	$18 - 4 = 14$	
3	either $14 - 9 = 5$	LET for event 3
	or $14 - 5 = 9$	
2	either $8 - 5 = 3$	
	or $5 - 3 = 2$	LET for event 2
1	$2 - 2 = 0$	

Critical Path

From the calculation of earliest and latest event times on a network it is possible to locate the sequence of activities that determines the overall project duration, and in the case of the example shown in figure 2.11 it can be seen that events 1–2–3–6–7–8 form such a sequence. The term critical path has been derived since this is the path that determines the project duration and hence any delay on any activity on that path is critical to the overall completion of the project. It is usually easy to locate the critical path by finding the events where the EET and LET are coincident, but this, while being a necessary condition, is not a sufficient one, since it is also necessary that the duration of any activity connecting two critical events shall be equal to the time difference between those two critical events. For example, if we added an activity from event 2 to event 6 of two days' duration in figure 2.11 it would start and finish at critical events but would not itself be a critical activity since its duration is less than the time available between those two events. If, however, such an additional activity 2–6 had a duration of twelve days, this would exactly equal the time available between events 2 and 6 and it would become critical, thereby giving a second critical path to the network, namely 1–2–6–7–8. It is perfectly possible that there be two parallel critical paths, but they must both be continuous from beginning to end of the network, it not being possible for a critical path to come to a dead end.

The identification of the critical path in a project network is of great importance, not simply because it determines the overall project duration, but it does set out a series of dates throughout the project that must be met individually if the overall completion date is to be met. In a realistic project network of say 500 activities, it is likely that not more than about 50 will lie on the critical path, and hence it is possible by concentrating control effort on these 50 critical activities that progress of the project may be more carefully monitored. If a project appears to be running behind schedule the common reaction is to increase effort all round, but in many cases this may be wasteful and it may in fact divert effort from the activities in which it is most needed. There is little point in increasing effort on activities that are not critical, but it is important that everything possible should be done to restore the project programme by concentrating additional effort on critical activities.

A further use of the critical path comes in the early stages of the project when the network plan is first prepared, and it is found perhaps that the calculated project-duration time is in excess of the time allowed for the completion of the work. The only way in which the project duration can be reduced is to reduce the length of the critical path either by reducing individual activity durations along the path or alternatively by changing the logical sequence of activities. If the network has been properly constructed it should not be possible to change the logic of the network unless some change in construction method is also involved. This could for example lead in a building project to a decision to reduce overall project duration by changing from *in situ* concrete to pre-cast concrete for part of the structure.

Float and Activity Dates

Calculation of the critical path in a network determines that certain activities, that is, the critical ones, must be carried out at a particular time if the project is to be completed on schedule; this implies that other activities do not need to be carried out at particular times and that there is some flexibility allowable in their timing. It is possible to see in figure 2.11 that while job 3–4 could be completed by day 10 at earliest it is not necessary to complete it until day 14 and there is therefore some spare time or *float* of 4 days on this activity. Similarly there is a possibility of varying the timing of activities 4–7, 2–5 and 5–6 in this particular network. There are various ways in which we may use float in a network, namely

(1) We may wish to extend the duration of *non*-critical activities so as to reduce the demands on labour and other resources.

(2) We may wish to delay the start of some activities in order to take advantage of better weather, more convenient construction conditions, or simply to allow some slackness in our management process.

(3) We may wish to sequence certain *non*-critical activities that require common resources for their completion; for example we may move one piece of equipment around from one activity to another in a convenient rather than a logical sequence.

In the case of small and medium networks where calculation is undertaken by manual methods there is really no need to use the more formal definition of float that has been evolved specially for use in connection with computer calculations. It is not the intention in this book to discuss in detail the various forms of float, since they are mostly related to computer calculation which is simply covered in outline form below. In the same way it is necessary to calculate earliest and latest start and finish dates for activities rather than calculate event times when a network is evaluated by computer. Again activity dates are not discussed in detail in this book but both these and float times can be studied in many of the more detailed books on the subject.[1,2,3,4,5,6]

Computer Calculation of Networks

Much useful work on the development of network planning methods has been undertaken in conjunction with computers, but this has had the unfortunate side-effect that many people have come to regard network planning methods and computers as being inseparable. As has been stated before it is quite straightforward to evaluate small and medium networks by manual means but there is no doubt that in larger networks computers can offer considerable advantages. The main advantage of a computer is that it is able to handle large amounts of data quickly and accurately and to present the output from the calculations in any form that is desired.

Network calculations consist essentially of a large number of arithmetical additions and subtractions, which while being tedious must at the same time be carried out accurately, since an error at any point in the calculation is carried right through and will continue to the end of the network. One single error in calculation can therefore invalidate the whole network. The computer is ideally suited to undertake these calculations both quickly and accurately and to print out the answers in a suitable form. It is, however, difficult to arrange for a computer to print out earliest and latest event times on to an actual network diagram although this can in fact be done. It is more common for the computer to print out the solution to the network in word and number form relating everything to activities rather than to events. It is usual to print out, for each activity, one line that gives the description of the activity, the numbers of the events at which it starts and finishes and four appropriate dates—namely the earliest start date, earliest finish date, latest start date and latest finish date. The meanings of these dates are fairly obvious; they are respectively the earliest and latest dates on which each activity can start and finish, as calculated from the earliest and latest event times. It is also usual to print out two types of float namely total float and free float. Again description of these terms in detail is beyond the scope of this chapter but, briefly, total float is the difference between the time available for an activity and the time required for its completion; and free float is that part of total float which does not cause any delay to subsequent activities. For further description of these and other forms of float see references 1 and 2.

No great skill is needed to use a computer for the evaluation of a network, since the calculation methods have all been worked out by computer specialists and put into a series of instructions to the machine. This set of instructions is known as a computer program, and virtually every computer will have available a relevant PERT or CPM program. The other input required to the computer is essentially a description of the network, but this description can be very simple indeed. The only information that the computer needs is a list of activities, stating for each activity the number of the event at which it starts, the number of the event at which it finishes and the activity duration. It is not *necessary* to input the job or activity description, but this is usually done so that the description may be printed out with the output information on the same piece of paper. In addition to the list of activities, it is usually necessary to add ancillary instructions regarding number of days per week worked, project start date, forms of output required, and so on. The usual form of output from a computer will be a printed list of activities repeating the input information but now adding the activity dates and float times. There are many variations to the output which can be produced and some of these are noted below.

(1) The activities can be listed in any sequence desired, perhaps in sequence of their earliest start dates, that is, first, all jobs or activities that can start on day 1, and then all those that can start on day 2, and so on. It may alternatively be desired that all critical activities be printed out first—these are of course those with zero float—and then follow the activities with one day float, then those with two days of float, and so on. Many other sequences can be specified.

(2) Many computers will store a complete dated calendar in their memory and can then translate from day numbers to actual calendar dates, and the output will then come in the form not of day number 1, day number 2, etc., but in actual calendar dates starting from the specified project commencement date.

(3) Another form of output is to actually print out the activities as a bar chart, usually using a row of Xs to indicate the timing of activities.

(4) It is sometimes possible to include in the activity description a code letter relating to the type of work involved in the activity. It is then possible in the print-out to separate activities of different types and perhaps produce separate print-outs for builder work, joiner work, plumber, etc. This is particularly advantageous where a large number of sub-contractors are involved in a project.

Updating and Monitoring

There is a considerable amount of work involved in setting up the data input to a computer for a network, but once this has been done it can be retained on magnetic tape and used again. This means that if it is required to make a few changes to a network, it is possible to put in the original data tape plus the small number of amendments and obtain a new calculation of the network very quickly. This brings out one of the major advantages of using a computer, namely that repeated calculations of the network are very rapid. It is usual in the early stages of calculating a project network that the implied project-completion date does not meet with the required contract-completion date. It will therefore be usual to make a number of attempts to bring the two into line either by changing the activity durations or the logic as described previously; with the help of a computer these subsequent calculations can be undertaken very rapidly. This facility is also very useful when the project has started and it is required to maintain surveillance over progress on the project. The usual practice is to update the computer data tape by indicating which activities have been completed by a particular stage and perhaps modifying the durations of forthcoming activities in the light of experience of the project to date. This process is known as updating or monitoring and on many projects is carried out on a weekly or monthly basis.

This whole process of updating and monitoring is of extreme importance irrespective of whether the computer is used or not. The whole essence of project control can be summarised in figure 1.2. First a plan is prepared and agreed and then, as work proceeds, measurement of progress in relation to the plan can be made. If, as is likely, the progress is found to have deviated from the plan, some form of control is needed to bring it back into line, and it is usually found necessary then to replan for the future. The plan–measure–control cycle will be continued throughout the project life. The important thing to remember in relation to time control of a project is that what has gone before a certain date in the project life is now of little consequence, and that effort must be concentrated on what lies ahead. It is always necessary to plan from *time now* until the end of the project. This regular updating, monitoring or replanning is of very great importance, but is often neglected because of the sheer amount of work involved in undertaking it. It is here perhaps that computers can be of the greatest help where updating is so easily completed.

Resource Allocation

It was stated early in this chapter that when a network is first prepared it should be carried out on the basis of unlimited resources since this was the only way in which the true logical dependency of activities could be set down. In reality of course no project is carried out with unlimited resources and in the completion of the project plan it is necessary to take account of resource limits. This can be undertaken by the utilisation of float time as was described in the section under that heading. A fairly simple approach to the problem of resource allocation is to prepare a network on the basis of unlimited resources and then produce from this the relevant bar-chart diagram showing for each activity both the earliest possible timing and the latest possible timing, as is done in figure 2.12 relating to the network shown in figure 2.11.

Suppose in this example that a tower crane is required for job 2–5 and also for job 3–4. On their earliest possible timings these two activities overlap and for them to be done in such a way it would require two tower-cranes on the site. This would seem to be an unreasonable approach and it would be more sensible to try to use only one crane. This could be done if activity 2–5 were to be completed by the end of day 7 and activity 3–4 delayed so that it started at the beginning of day 8 thereby finishing at the end of day 12. This type of manipulation can readily be done manually on a bar chart and in practice this is often proved to be the most effective method. On large projects the procedure for doing this would become extremely complex and again use can be made of the computer to carry out these calculations.

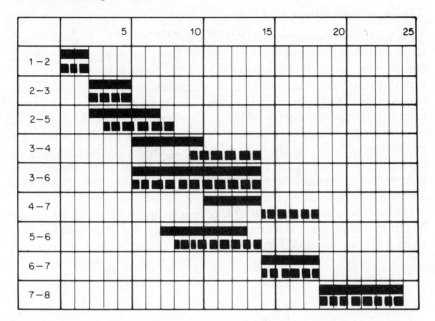

Figure 2.12   Bar-chart representation of network in figure 2.11

Resource-allocation programs are rather complex and again outside the scope of this book, but it is sufficient to say that they exist and that they can be utilised to carry out the smoothing of up to 100 separate resources on a major project.

Summary of the Procedure for Planning and Control by Networks

The sequence of operations involved in project management by networks can be summarised as follows.

(1) Prepare an arrow diagram showing the *logic* of the project without reference to time or resource limitations.

(2) Estimate activity *durations* and calculate earliest and latest event times and project overall duration.

(3) *Check*, and adjust if necessary, logic and durations so that the overall projects fits the time available, if this is possible.

(4) Use float to carry out *resource smoothing* and produce a working plan.

(5) Start project and *monitor* regularly to maintain on-going control of the project.

(6) At each updating stage *replan* the balance of work outstanding.

Time–Cost Interactions

Most construction managers will be well aware that 'time costs money' and it follows from this concept that time can be bought. It is usually possible in a construction project to reduce the overall duration by the expenditure of further money. Since a project can only be shortened by a reduction in the length of its critical path it is useful to examine the critical path to see if there are any activities on it that would be reduced in duration by the application of further effort, usually at some cost. There exist methods whereby it is possible to calculate the cheapest way of reducing the length of the critical path, and one of these is discussed below. It is important to note the principle that by the application of further effort to critical activities the project duration may be reduced.

For any single activity it is possible to represent the relationship between its direct cost of plant and labour (but excluding materials) and the time taken for its completion. Figure 2.13 shows a typical relationship in which

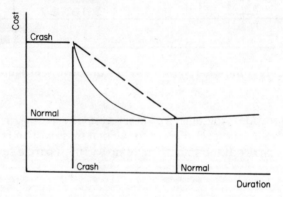

Figure 2.13   Time–cost relationship for a single activity

the cost of the activity is at a minimum if the work is completed in what is thought to be *normal* time. This so-called *normal* cost should be minimal since it should normally be the intention of site managers to carry out each activity at least cost. If the activity takes longer than this normal time it is likely to cost more money, probably because work is being done inefficiently, and if it is required to complete the activity in less than normal time it may be necessary to expend further effort or money to achieve a reduction in time. It is usually possible on any individual activity to continue to reduce its duration by additional labour, overtime working, double-shift work or special bonus, until the point is reached where it just cannot be done any more quickly, and this is called the *crash* time. Figure 2.13 indicates that there is a smooth curve representing the relationship between cost and duration but in practice it may be the case that the line consists of a series of steps or may simply be a series of single points, depending upon the nature of the work involved. In order to simplify the calculation procedure it is usual to represent the relationship by a straight line drawn between the crash point and the normal point as indicated by the dashed line in figure 2.13. This relationship can then be expressed as algebraically as follows

cost of one time unit reduction in duration

$$= \frac{\text{crash cost} - \text{normal cost}}{\text{normal duration} - \text{crash duration}}$$

This straight-line approximation is adequate for most purposes and gives much easier computation.

For the network illustrated in figure 2.14 the following costs and durations apply as set out in table 2.2.

Table 2.2   Time–cost data for figure 2.14

Activity	Crash cost	Normal cost	Normal duration	Crash duration	Cost per unit time reduction
1–2	£60	£40	4	3	£20
2–3	£75	£30	5	2	£15
2–4	£60	£36	7	5	£12
3–4	£100	£50	4	2	£25

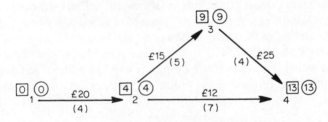

Figure 2.14   Network with time–cost interactions

The objective is to reduce the duration of the project by one time unit with the least possible expenditure. Examination of the network in figure 2.14 will show that the critical path passes through activities 1–2, 2–3 and 3–4, and in order to reduce the project duration it is necessary to reduce the duration of the critical path. The cheapest way of achieving a reduction of one time unit is to reduce the duration of activity 2–3 to 4 units at a cost of £15. This is indicated in table 2.3 along with the calculation of subsequent further reductions of one time unit.

Table 2.3   Calculation of cost of successive time reductions

Time unit	Activity reduced	Cost of reduction
1st	2–3	£15
2nd	2–3	£15 job 2–4 now also critical
3rd	1–2	£20 job 1–2 now crashed
4th	2–3 and 2–4	£15 + £12 job 2–3 now crashed
5th	2–4 and 3–4	£12 + £25 job 2–4 now crashed
		£114 total cost of reduction

Table 2.3 shows that the total cost of reducing the project duration by 5 time units is £114. If the project overhead costs are assessed at £25 per time unit it would appear that it is just worthwhile to expend £114 in order to achieve the overhead saving of 5 × £25 = £125. However, careful examination of table 2.3 will show that the maximum overall financial saving can be achieved by reducing the project duration by 3 time units only, since each of the first three time units will have cost less than £25 to achieve. In order to obtain the fourth and fifth unit reductions in project duration it is necessary to spend £27 and £37, which is in excess of the saving of £25 on overheads.

It can be appreciated that in a complex network the above calculation would be extremely tedious, and it is normal to carry out calculations of this type by computer. It may not be necessary to carry out this detailed assessment in many practical projects because it is fairly simple to identify the activities that offer the opportunity of time reduction by the expenditure of additional resources. A very clear example of this is described in detail in chapter 3, relating to the construction of a road bridge over a railway. This illustrates well the principle that it is often possible to identify a critical activity on which the expenditure of money will bring about a reduction in the overall project duration, and produce a consequent financial saving.

Non-linear Time–Cost Relationships

The linear assumption of the relationship between cost and duration may frequently not be realistic. In some cases there could be only two points to consider, the normal time and the crash time; for example, digging a small hole could either be done by a man in two days or by a machine in one hour, and intermediate times may be meaningless. In other cases the relationship may be far from linear, especially where to break through to a shorter duration involves a new method with a significantly increased cost. The simple arithmetical method described above becomes unworkable in these cases, but it is possible to use a dynamic programming approach, based on the principles discussed in chapter 9. The method is too specialised for inclusion here, but it has been described elsewhere by the author.[7] The principle of the method is somewhat similar to that described above and it proceeds by asking the question 'If a sum of £10 is to be expended in accelerating an activity, where will it have greatest overall effect?'. The question is repeated successively with £10 increments either until the required time reduction has been achieved, or until no further reduction is possible.

Planning of Repetitive Operations

In a limited range of projects whole sequences of activities are repeated, for example, in multi-storey flat building, or large estates of similar houses. An overall network analysis is not necessarily the most appropriate planning method in repetitive situations, and it is often preferable to use a network to study in detail only one cycle of the project, thereby establishing a pattern that may be followed in all the subsequent cycles. The planning of effective resource use is essential in this type of project, and many specific techniques have been evolved to handle this situation. One of the most useful methods is 'line of balance', which is a form of linked bar-chart.[8] This does not necessarily supercede the use of networks, but when the two methods are used in a complementary way they can together provide a powerful planning tool.

References

1. K. G. Lockyer, *An Introduction to Critical Path Analysis*, Pitman, London, 1970.
2. K. M. Smith, *A Practical Guide to Network Planning*, British Institute of Management, London, 1965.
3. K. G. McLaren and E. L. Buesnel, *Network Analysis in Project Management*, Cassell, London, 1969.
4. R. D. Archibald and R. L. Villoria, *Network Management Systems*, Wiley, New York, 1967.
5. A. T. Peart, *Design of Project Management Systems and Records*, Gower, London, 1971.
6. J. J. Moder and C. R. Phillips, *Project Management with CPM and PERT*, 2nd edition, Van Nostrand Reinhold, New York, 1970.
7. J. F. Woodward, *Time/Cost Interaction in Project Management*, Internet Conference, Paris, 1974.
8. N. P. Lumsden, *The line-of-balance Method*, Pergamon, Oxford, 1972.

3 Construction Planning

In the whole field of the construction industry the word 'planning' can have a number of different meanings. In the macro sense, such as in urban or regional planning, the subject concerns major decisions on what types of building shall be constructed and what their general form will be. Within the engineers' terminology this may more truly be said to be a design phase. When the design of a building is complete and the drawings and other documents are passed to a contractor he will be concerned with his own aspect of planning, namely considerations of the methods and means he will use to execute the works already designed. This is the work traditionally carried out in the planning department of contractors' organisations. The third use of the word 'planning' relates to the preparation of a time-schedule for the completion of the works whether by bar-chart or network methods, and is perhaps more truly referred to as programming.

These three separate aspects of planning may be briefly stated as follows.

What to do
How to do it
When it is to be done

The first of these covers the whole range of design—social, architectural, and structural—and some aspects of this part of construction are covered in other chapters of this book. The third area concerning programming is discussed in chapter 2 particularly with respect to network methods, but reference is made to other programming methods in that chapter. The essence of the present chapter is to examine in detail a few possible applications of quantitative methods in decisions relating to actual construction methods to be utilised on site.

It may seem unrealistic to some readers to separate the design phase from the construction phase by an absolute division, and it is certainly not the objective of this book to suggest that such a division should necessarily be made. In practice it is certainly true that designers take account of the possibility of different construction methods when designing a particular structure. It is the case however that there is in many ways an effective separation, especially when design work is complete before the contractor for a particular project has been named. Even where the contractor is involved in the design such as in the package deal design-and-build type of contract, it is possible that the two functions are undertaken by separate departments without full advantage being taken of the fact that they are both within the same company. In this respect the construction industry has much to learn from the manufacturing industry; for example motor vehicles, where the methods of manufacture exert a very large influence on the design of the final product. While accepting that there is not necessarily a sharp dividing line between design and construction, it is none the less convenient to consider that there are design problems and construction problems which may be looked at separately. It is the latter which forms the subject of this chapter.

There is similarly no sharp dividing line between taking decisions on construction methods and setting out a timed programme of work. One of the most important aspects of construction planning is to consider the time required, and it is of course well known that time and cost will interact in a project. As before, it is convenient to separate the programming techniques such as network analysis and bar charts from the study of the ways in which we may examine alternative construction methods for any particular project. Again it is the latter which forms the subject of this chapter.

The Need for Planning

When management techniques are discussed with industrialists a typical reaction is 'These techniques may work well in other industries but our situation is different, our problems are much more complicated and the techniques just would not work for us'. It is perhaps easy to understand how this view should arise but in most cases it is a fallacious one. It is certainly true that each industry is *different* from the next and has to cater for special problems of its own. It is also true that some techniques are only applicable to certain industries, but the view that one's own industry is more difficult than the next man's is of necessity a fallacy. Problems will be different but not necessarily more difficult. The popular misconception may derive from an over-zealousness on the part of teachers of management science who try to adapt problems to fit their own pet techniques. It is important that each industry should select only such techniques that are felt to offer assistance to the manager, but it is possible that the general approach of the management scientist using his techniques of search, quantification and comparison can offer valuable assistance in any industry.

In the context of this general view it is now valuable to look at the part played by planning of construction methods. Some of the features which typify construction work and make it different from other industries are as follows.

(1) Most construction projects are unique; an exact precedent of any particular project probably does not exist and it is unlikely that successive identical projects will be built. While each project as a whole is unique it is

equally true that most elemental parts of the project will not be unique and will have been undertaken elsewhere before, and are likely to be repeated again.

(2) Most projects are multi-elemental; even a small construction project will have many thousands of separate operations to be completed, and in large projects the number of operations could well run into millions. The sequence in which these elements fit together in a project is complex and most of the elements will interact.

(3) The structure of the industry is highly fragmented, it being common for the following separate organisations to be involved in any one particular project: architect, structural engineer, services engineer, main contractor, sub-contractors for each specialisation, surveyors, materials suppliers in great number. There can easily be more than a hundred separate organisations involved in the construction of an average project.

The fact that most projects are unique means that the trial-and-error approach to construction is not valid. In manufacturing industry where a large number of identical units are to be made it is perfectly reasonable to try out one method of production, examine it in detail, and improve the method for the second and subsequent units. It is possible to do this several times until the best method is found and it is quite reasonable to spread the cost of the trial-and-error work over a large production volume. This is just not possible in construction where each project is unique, but it may be reasonable to use this approach on elemental parts of the project. For example, much can be learnt about the fixing of shuttering, or laying of bricks, on one site that is applicable to the next site, but it must always be remembered that while conditions are broadly the same on two sites, there may be other factors which render the two situations non-identical. The fact that it is not possible to approach a construction project on a trial-and-error basis means that some form of planning is essential. This general principle is embodied in the popular phrases 'Plan ahead' and 'Look before you leap'. Some of the major difficulties of the construction industry compared with other industries relate to the multi-elemental nature of the work and the fact that these highly inter-connected elements are the responsibility of many different organisations. If these elements are to fit together in an efficient way it is essential that a whole project be properly planned and programmed.

Major Factors in Construction Planning

It has been stated above that it is essential to carry out some form of planning and programming before a construction project commences. It is important to do this in order to avoid, as far as possible, difficult situations that will arise if each individual organisation attempts to tackle its part of the work in an independent rather than an integrated way. This is essentially an argument in favour of programming in the sense that it was used in relation to network methods in chapter 2. However, it also makes the case for planning, that is pre-decision on the methods of construction to be used for a particular project. It is barely adequate to predetermine only a single method of approaching each part of the project, since in most cases there will be the possibility of using more than one method. It is therefore useful in each situation to compare alternative methods available so that the most appropriate one may be selected. The basis of comparison and method of selection requires careful consideration of the following factors.

(1) Cost is the factor that predominates in the mind of the contractor and is usually given overriding importance in the comparison of alternatives.

(2) Time is an important factor, especially where a reduction in time offers a financial saving. Time and cost will of course frequently interact and this subject is discussed more fully later in this chapter.

(3) Resource availability may determine choice of construction method, for example the ready availability of specialised items of plant or specialised skills may offer an advantage to a contractor in a particular situation.

(4) The quality of the finished work may often depend upon the method used, and it is common practice to use the cheapest or quickest method that will just satisfy the quality for the project. This is far from being a criticism of cheap or quick methods of construction; on the contrary there is little point in using expensive or time-consuming methods to produce quality that is considerably better than that specified. This simply represents a waste of time and money.

(5) Safety is a factor that should predominate in decisions on methods of construction but unfortunately the safest methods are often the most expensive ones. Safety is not quite in the same category as quality in that it is not easy to define a standard of safety that is just good enough, since it will always be possible to argue that an even more safe method is justifiable. The only guide that can be given is that methods should in all respects comply with the appropriate safety regulations for construction sites and that all equipment should be tested as required by these regulations. It may be of some consolation to the planner that construction-site accidents are less frequently the result of wrong decisions on the type of construction method to be used, and more frequently due to the method decided upon not being properly executed.

Comparison of Alternatives

It has been stated above that it is the task of the construction planner not simply to set down *one* method of completing a construction project but to compare alternative methods applicable to various parts of the project. This should be undertaken methodically in the following sequence.

(1) The selection of operations that merit detailed study. It is commonly found in industry that a small proportion of the total number of different products of a company represent a large proportion of the company's turnover and profit; conversely a large proportion of its products (the less popular ones) are responsible for only a small proportion of its turnover and profit. A similar division arises on a typical construction project and examination of a bill of quantities may well show that 20 per cent of the bill items (the biggest ones) add up to about 80 per cent of the contract value. The remaining 80 per cent of items, mostly the small ones, will only contribute about 20 per cent to the contract sum. While these actual percentage figures will obviously vary from contract to contract, the general tendency is common and in some specialised types of work, for example pipe lines, the concentration of value in a small number of items might be even more marked, say 90 per cent of the value contained within 10 per cent of the items. It therefore makes sense when selecting operations for careful scrutiny to consider first those which represent a major part of the construction cost, since any unit cost savings that may be achieved by an improved construction method will offer a larger overall saving. This probably means that planning efforts should be concentrated on basic activities such as excavation, large areas of shuttering, brickwork, placing of concrete, rather than more specialised small items.

(2) The generation of ideas for alternative construction methods is an important positive step. It is not adequate to consider only the most obvious method for any particular operation and simply compare it with one alternative as a check. Once a particular operation has been identified as being worthy of detailed study, a positive step should be taken to write down a number of different methods for its completion. Some of these alternatives may at first appear to be inappropriate but it is none the less worth writing them down. It is not necessary to evaluate all alternatives in detail and a sequential screening process may be used. Some of the screens may take relatively little time and money to apply and may therefore be applied first. For example the first screen may simply ask 'Does the company have access to the technical resources required to use the proposed method?'. This question might easily eliminate one or more of the alternatives proposed. The second question might be 'Is the proposed method likely to be acceptable to the architect or engineer?'. Again this question can probably be easily answered, thereby eliminating further alternatives. Questions of quality and safety may in turn be posed making further eliminations from the list, leaving perhaps a small number of alternatives to be fully evaluated.

(3) The evaluation of the alternatives left from the above screening process should be carried out to determine what the cost of the proposed method would be and how much time it will take. The cost will almost always be a relevant factor to be considered and should be worked out with reasonable accuracy. It is important to note, however, that it is not necessary to have absolute accuracy for the cost estimate since it is only to be used as a

basis of comparison between two or more alternative methods, and is not an absolute price that is being used for the purpose of quotation. This of course only applies if alternatives are being considered after a tender has been submitted; if the planning alternatives are being studied during the preparation of the tender then of course the cost estimate of the method selected will be used as part of the total cost build-up for the whole project, and must therefore be appropriately accurate.

The time estimate for a proposed construction method will often but not always be of importance. If the overall project programme allows plenty of time for a particular operation then there will be no advantage in selecting the quickest method of completing the work. If on the other hand the overall project duration is affected directly by the time required to carry out the operation then time will be of the utmost importance and may even appear to be of disproportionate importance to that operation. There is an intermediate situation where, while an operation may not be critical to the project completion, it may constrain other work if it occupies a longer period, and in this situation the time required is of moderate importance. The accuracy with which the time for an operation is estimated will depend upon the importance attached to it. Clearly if there is plenty of time available in the programme then the accuracy of time assessment can be poor, but if the operation is critical to the project completion date then its time estimate should be as accurate as possible.

The above approach can best be illustrated by the following simple example.

Alternative Methods of Shuttering Ground Beams

This example concerns a steel-framed single-storey industrial building with brick walls. The foundations consist of a series of individual pad bases for the columns linked together by ground beams to support the perimeter brick wall. There are similar ground beams supporting internal brick partitions. While it is not one of the very large items in the bill of quantities the shuttering of the foundations of ground beams is a significant item and is thought to offer the possibility of saving of time and/or money. A number of alternative methods for shuttering the side walls of the beams and bases are written down as follows.

(1) Purpose-made plywood forms made off-site and simply fixed by the site carpenters or joiners.

(2) Rough-sawn boarding cut and fixed directly by site labour.

(3) Proprietory steel forms.

(4) Single-skin brick wall permanent shutter.

(5) Corrugated-iron sheets (probably second-hand) left in as permanent shutter.

(6) Corrugated-asbestos sheets left in as permanent shutter.

(7) Concrete poured directly against the sides of a slightly over-sized excavation.

These seven alternatives are set out in table 3.1 which also lists the qualitative screens that may be sequentially applied. It can be seen from this table that of the seven alternatives only three pass the screens, and it is therefore only necessary to evaluate the cost and time involved of these three.

Table 3.1   Alternative foundation shutters—qualitative screens

	Acceptable to architect?	Feasible?	Adequate quality?	Safety?
(1) Plywood panels	yes	yes	yes	yes
(2) Rough-sawn boards	yes	yes	yes	yes
(3) Steel form	yes	*difficult	—	—
(4) Brick skin	yes	yes	yes	†doubtful
(5) Corrugated iron	no	—	—	—
(6) Corrugated asbestos	yes	yes	yes	yes
(7) Trench sides	no	—	—	—

* The variation between bay lengths would make the use of steel shutters unacceptable because of difficulty in cutting and making up.

† Brick-skin shutters are only strong enough for shallow beams. In this example it is thought that there is a risk of the brickwork collapsing under the pressure of wet concrete.

Comparison of costs and time is set out in table 3.2. It can be seen from this that the use of sawn boards is marginally cheaper than plywood forms and considerably cheaper than asbestos sheets. This would probably mean that if cost were the only criterion for decision it would be decided to use the sawn board, but it should be noted that since the two figures are close the decision between them would be fairly sensitive to changes in the component cost. In the lower part of table 3.2 it is shown that if the operation is required to be carried out in minimum possible time, then the use of asbestos sheets becomes preferable. In order to obtain full advantage of the plywood shutters time must be given for ten uses of the forms to be obtained, namely 30 days. If sawn boards are used with a smaller number of uses the whole operation can be completed in a shorter time. If time is extremely important the use of asbestos sheeting as permanent shutter would enable the operation time to be reduced to three days, but at considerably greater expense. It is unlikely that it would be possible to obtain sufficient resources to fix all the asbestos sheeting in a period as short as three days and it is thought that this method, while potentially offering the shortest operation time, could not be adopted. Another consideration, however, would be the trade resources required for the operation and while both plywood and sawn-board shutters require five man-hours per unit for fixing it can be seen that the proportion of tradesmen-hours is much lower in the case of plywood shutters. This would mean that in a situation where tradesmen were in short supply it would be preferable to use the plywood shutters since they make more use of unskilled labour. It is also worth noting that if it were decided to use the asbestos sheets in order to carry out the operation in the absolute minimum time of three days, then it would be necessary to have a labour force of twenty carpenters and forty labourers—almost impossible requirements in many situations.

Table 3.2 Alternative foundation shutters—cost and time comparison

Cost per unit of shutter	Plywood	Sawn board	Asbestos
Material cost	£17	£4	£3
Off-site making cost	£15	—	—
On-site fixing cost per use	£8	£9·5	£10
Number of uses	8	3	1
Residual value	£4	—	—
Total cost per use	£(17+15−4)/(8+8) = £11·50	£(4/3)+9·5 = £10·83	£3+10 = £13
Cycle time (days)	3	3	3
Minimum operation time (days)	3×10 uses = 30	3×3 uses = 9	3×1 use = 3
Number of units	240	240	240
Carpenter-hours per unit	1	4	2
Labourer-hours per unit	4	1	4
Labour required to complete in minimum time:			
carpenters	1	13·3	20
labourers	4	3·3	40

The above very simple example illustrates the procedure that may be followed in examination of alternative methods of construction. Many other examples could be described but inclusion of these in detail would be tedious and largely pointless. However in order to give an idea of typical situations the following possible applications are briefly outlined.

The Construction of a Deep Culvert

The construction of a culvert of, say, 2 m diameter could broadly be tackled in any one of three ways as follows.

(1) Open excavation with the sides of the trench standing at an acceptable batter. This method would involve a large volume of excavation and subsequent back-filling, would occupy a large amount of surface area and would therefore not be possible if existing buildings or other surface features had to remain undisturbed. It also might lead to some access difficulties since cranes and other plant would not be able to approach close to the line of the culvert. It would, however, be possible to open up a considerable length of culvert at one time and enable the work to be completed relatively quickly. The engineer would have to be satisfied that the large volume of disturbed material could be sufficiently compacted upon being replaced, and the slope of the side of the excavation would have to be adequate to eliminate the risk of a slip.

(2) The first alternative method of construction would be to excavate the trench or the culvert between sheet piling and thereby greatly reduce the volume of excavation involved. There would of course be the additional cost of sheet piling to be covered, but it would be possible to get closer approach to the line of the culvert. Access would be somewhat impaired by any shoring necessary to keep the piling in position, but as in the case of the open cut it would be possible to work on a considerable length of the culvert at one time. There would of course be less material to excavate and back-fill and it is likely that it would be easier to ensure safety by adequate shoring of the piling than would be the case in the open-cut method.

(3) Tunnelling would provide another alternative, and would of course require a totally different technique. It is the method that would minimise the amount of material to be excavated, but the method of excavation would be more costly. The shuttering method would have to be quite different and concrete placing would be rather more difficult. It is the method that offers least interference to the ground surface and may therefore be preferred by the client, but is likely to be the slowest method since it can only be tackled from two points compared with either of the open methods, which could be tackled from several points simultaneously.

Variations on these three basic alternatives can be made by the use of pre-cast rather than *in situ* concrete, thereby giving at least six alternative schemes. It would be necessary to make a detailed evaluation of the costs involved in each of these schemes together with an estimate of the time required for their completion. It is likely that the first alternative would be preferred at shallow depths, the second alternative at intermediate depths and the third alternative at greater depths; but it is not easy to determine the depths at which preference changes from one alternative to another. This would have to be carried out by detailed calculation, the result of which could show that on a particular contract all three methods might be justified on different parts of the culvert length depending upon the depth at each point. It would then be a matter of deciding whether to use one consistent method throughout or to use some combination of two or more methods.

Access to the External Facing of Buildings

Access to the external face of buildings under construction presents a familiar problem to managers in the building industry. It is again useful to list a number of alternative means of providing access and these can first be screened for acceptability, practicability and safety before cost and time estimates are made. The following list of alternatives is by no means exhaustive but does indicate some of the methods possible.

(1) Free-standing conventional tubular-steel scaffold.

(2) Proprietory interlocking free-standing scaffold.

(3) A boat or cradle suspended from the roof.

(4) Suspended scaffold fixed at each floor as the work proceeds and raised by crane.

(5) Movable tower scaffold carrying out the work in verticle strips.

(6) Working entirely from within the building.

Timber-frame House Construction

The use of timber-frame construction for house building is not yet widespread in Britain but is becoming more popular. There are four broad categories of construction method that may be used as follows. (It is appreciated that in this case the construction method must be incorporated into the design, and it is not entirely at the discretion of the building erector as to which method he should use.)

(1) Building up piece by piece using timber cut from random lengths delivered to the site.

(2) Building up on-site wall panels made up from precision pre-cut timber lengths delivered from a workshop. The panels are then erected to form the finished structure.

(3) Factory-built panels delivered to the site and then erected. These panels are generally pre-glazed but do not carry the external cladding.

(4) Factory-built room-sized units complete with all services and most internal and external finishes. These are then simply placed together on site on pre-prepared foundations and joints between them made good.

The choice between these alternatives has presented considerable problems to housing contractors in Britain. Many of them have found that the advantages potentially offered by the factory methods of production have not fully materialised due partly to technical difficulties and partly to the lack of adequate production throughput. Detailed discussion of these alternatives is not possible in this chapter but it is worth noting in passing that in Canada, often regarded as the origin of the timber home, it is usually method 2 above that is applied in preference to the more capital-intensive factory techniques.

Large Wall Shutters

A reinforced-concrete wall 2 m high is to be constructed, and two alternative types of wall shutter are considered as in figure 3.1. Type A consists of standard plywood panels with timber battens and loose soldiers at 400 mm centres. Each soldier is fixed with two through ties. Type B is essentially similar but with the addition of pre-fixed steel channel walings, which permit a big reduction in the number of through ties, it being necessary to place ties at only each fifth soldier. Comparison of costs is as follows in table 3.3 for a panel shutter 2 m high by 4 m long.

Table 3.3   Cost comparison of shutter types

	Type A	Type B
Making cost of plywood panels	£24	£24
Making cost of soldiers	£12	£15
Cost of walings and fixing	0	£45
Cost of moving panels	£1 (by hand)⎫	£2 (by crane)
Cost of moving soldiers	£2 (by hand)⎭	
Cost of fixing shutters to line	£15	£10
Cost of striking	£3	£2
Cost of ties (per use)	£10	£5
Total 'make' cost	£36	£84
Total 'fix' cost per use	£31	£19

In a case where a very large number of uses are possible, for example in the construction of a high wall in several lifts, it may be justified to expend a large sum of money on the original making cost, if the fixing cost can be reduced. This can be done in many ways

(1) an increase in the panel size;
(2) increased strength permits fewer fixings;
(3) increased rigidity will simplify the alignment.

Figure 3.2 shows how the total shutter cost varies with the number of uses, indicating that if more than four uses can be obtained it is preferable to use the stronger shutter, even though it costs more to make.

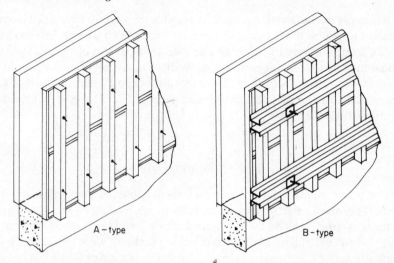

Figure 3.1   Alternative wall shutters

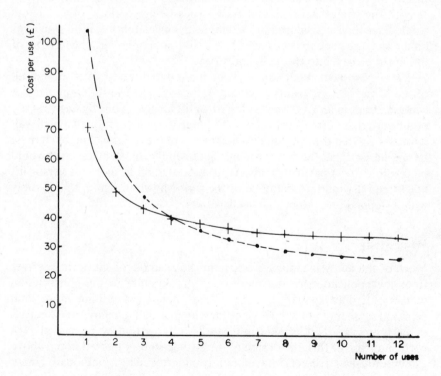

Figure 3.2   Cost per use of different shutter types

Referring to the above example, if the shutter were to be used, say, 100 times, it might be possible to use a steel form (type C) costing £800 to buy, with a fixing cost per use of £12. The total cost per use of type-C shutters would then be £800/100 + £12 = £20. With as many as 100 uses it would be necessary to reface A- or B-type shutters after each 10 uses at a cost of £12 per reface, and to replace the panels completely after 50 uses. The total cost per use would then be

$$\text{A-type } [36 + (8 \times 12) + 24]/100 + 31 = £32 \cdot 56$$
$$\text{B-type } [84 + (8 \times 12) + 24]/100 + 19 = £21 \cdot 04$$
$$\text{C-type } \qquad\qquad\quad 800/100 + 12 = £20$$

This would indicate that it is just preferable to use the steel type C, but the cost difference is relatively small, and the decision could be highly sensitive to the cost data. It can be seen that the decision could be changed if the fixing cost of B were to fall by £1·04 or if the fixing cost of C were to rise by £1·04. Similarly an increase in purchase price of the steel shutter C of £104 (13 per cent) would change the decision, but the decision is not at all sensitive to the accuracy of the making cost of the B-type. (It would be necessary to reduce this to about 40 per cent of its estimated cost to change the decision.)

With the choice between B- and C-types being so close, other factors would have to be considered; for example, the availability of lifting facilities on the site, the need to cut or modify the shutter at openings, the number of fittings to be cast into the wall, and so on.

In this example comparison has been limited to three types of shutter, all of the same size. In reality it would be possible to use different sizes of shutter, either in terms of shutter length or lift height. In one extensive study of the variation of total cost with lift height in the construction of a high wall it was shown that the optimum lift height was of the order of 2 m, but it must be emphasised that this is not a magic figure applicable in all cases. It would be possible, for a particular project, to plot a graph of unit cost versus lift height and it would probably be of the form shown in figure 3.3, possibly with a stepped line rather than a curve.

Work Study

Much of the analysis indicated above in the example of shuttering derives from the application of the methods of work study to the construction industry. It may appear that the above analysis is nothing more than common sense, but none the less the fact remains that in many areas analysis of this type was not undertaken until work-study methods were applied. As a particular case of this, the design of the B-type shutter quoted in the above example arose as a direct result of a study undertaken by a methods engineer on a construction site. He observed that a significant proportion of the labour time required for the placing and fixing of the shutter was taken up in

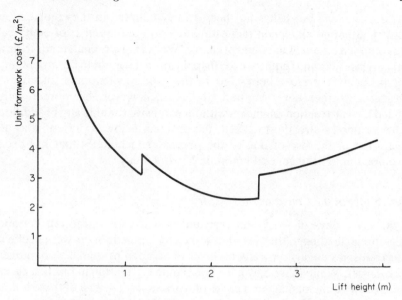

Figure 3.3   Unit formwork cost as a function of lift height

the attachment of the through tie bolts. He concluded that anything that could be done to reduce the number of bolts to be fixed on any particular shutter was worthy of consideration. This thought was then extended to the principle that if it were possible to make even a small reduction in the time required for repetitive operations, by the use of a stronger or stiffer shutter panel, then it was possible that the increased cost of such a panel could be justified. The same study was extended to look at a very much more sophisticated shutter using very large steel panels and its own self-erecting system. While in theory this offered potential savings by having fewer fixings and being a very rigid shutter, it was found that the lifting system led to difficulties of alignment leading to increased labour costs which more than off-set any potential saving that the system offered.

Work study involves two broadly different types of technique, namely those of method study and those of work measurement. Method study as the phrase implies is concerned with an evaluation of the methods being used to carry out a particular operation, and examination of the alternatives available. It is essentially therefore the sort of thing that has already been discussed in this chapter. Work measurement on the other hand is largely concerned with the measurement of work in time units of the individual work elements in an operation. It is often used in a production-engineering context to provide basic data for the preparation either of incentive bonus schemes or for the construction of time and cost estimates for the manufacture of a new product. It could therefore be of value in the preparation of tender estimates, but it has not been widely adopted in the construction industry although many contract estimators do use a number of empirical rules founded on much the same principle. Work measurement tends to be a fairly precise task and requires expert application if it is not to be misleading. It is probably therefore better left to the expert practitioner and is not considered further here. Method study can, however, be more readily applied by construction engineers without extensive training and in addition to the examples already discussed, the problem below is presented as an illustration of the use of one of the particular techniques that has been developed under the general umbrella of work study.

Work Study in the Construction Industry

In the early days of work study an unfortunate misconception became wide-spread leading to the view that work-study practitioners were in effect industrial spies engaged by an employer in an attempt to make operatives work harder. While there may have been some foundation to this fear there remain other quite distinct objectives of work study. The first of these is that work done must be effective in terms of both quality and quantity. It is desirable to eliminate wasted work such as double-handling, or even multiple handling as often occurs with bricks on a building site. There are many other forms of unnecessary work in building especially with respect to finishes; for example, a steel float finish to a layer of blinding concrete; shuttered concrete surfaces are often rubbed down prior to being covered by a skin of brickwork or patent cladding; frequently finishing work is completed and is spoiled by subsequent operations and consequently has to be repaired or even redone. In all of these examples work is being done ineffectively, and similar waste may occur if work is being done by the wrong person, for example labourers' work being done by skilled highly paid craftsmen. The key to the elimination of wastes of this sort is to develop a questioning attitude, and this may be summarised by the following simple questions relating to any particular operation.

> What is being achieved?
> Why is it being done?
> Where is it being done?
> Who is doing it?
> How is it being carried out?

If each operation in a construction project is questioned in this way the answers may well lead to suggested alternative ways of carrying out the work. Most of the questions, answers and alternative ideas are well within the competence of construction managers and others with experience of working on sites. They are not all followed through in detail here therefore, but as an example the placing of concrete illustrates the *how* problem. There are many good references covering the principles and application of work study, some of them relating to the construction industry.[1,2,3]

Use of Method Study in Concrete Placing

In the very simple case of concrete being poured in the foundations of a small building, a common method of operation is for a small concrete-mixer to be operated by one man, another man wheels hand-barrows full of concrete from the mixer to the place where the concrete is to be deposited, and a third man spreads, compacts and finishes the concrete. A simple observation of such a method may show that the man operating the mixer is only effectively working for a small fraction of the available time, as is the man spreading and compacting the concrete. The unfortunate man wheeling the barrows, however, may find that he is working flat out and is constantly pressed both by the man at the mixer and by the man spreading the concrete. Under these conditions it does not take an expert work-study man to suggest that there should be a second man with a wheel barrow, and perhaps that there may be spare barrows in use so that there is always an empty one waiting at the mixer ready to be filled as soon as the concrete batch has been mixed. The pace of such a set-up is largely geared to the rate at which concrete may be mixed and of course the output of the whole system cannot exceed the mixing rate. It may not be thought worth while to carry out a detailed study of such a simple operation but if a similar problem on a larger scale should arise then detailed consideration may well be justified. The following example illustrates just such a case.

On a large construction project concrete is mixed in a central batching-plant and then distributed to various parts of the site by truck. The trucks discharge either directly into the shutters, in the case of foundation structures, or into large skips for work above ground level. These skips are then hoisted by tower crane and concrete placed in the formwork. At one stage of the work where large volumes of concrete are being poured at a high level it is found that a substantial proportion of the time available on the cranes each day is being utilised for the placing of concrete. It is decided to investigate whether it would be possible to speed up the rate of placing of concrete using the initial method, or alternatively to install a pumping system and relieve the cranes completely of concrete work.

The first step is to make a detailed study of the existing method, taking care to ensure that the mere fact that the work was being observed in detail did not modify the way in which the work was being done. If the purpose of the observation is not explained to the work-force the fact that they are being observed may encourage them to either work faster than normal or slower than normal, depending upon their attitude. It is normal practice to observe several cycles of such an operation and to ignore particularly fast or particularly slow cycles, and concentrate attention on less extreme values.

Other methods include random sampling which, taken over a long period, will even out any irregularities or inconsistencies in the observation.

In the problem under consideration the result of the observations could be shown in the *multiple activity chart* given in figure 3.4. This shows that there are six separate units involved in the concrete-placing operation, namely the

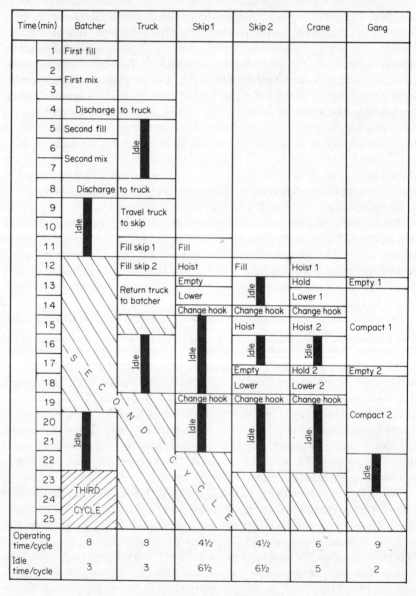

Time (min)	Batcher	Truck	Skip 1	Skip 2	Crane	Gang
1	First fill					
2	First mix					
3						
4	Discharge	to truck				
5	Second fill					
6	Second mix	Idle				
7						
8	Discharge	to truck				
9	Idle	Travel truck to skip				
10						
11		Fill skip 1	Fill			
12		Fill skip 2	Hoist	Fill	Hoist 1	
13		Return truck to batcher	Empty	Idle	Hold	Empty 1
14			Lower		Lower 1	
			Change hook	Change hook	Change hook	
15			Hoist	Hoist 2	Compact 1	
16	SECOND	Idle	Idle	Idle	Idle	
17						
				Empty	Hold 2	Empty 2
18				Lower	Lower 2	
19			Change hook	Change hook	Change hook	
20			Idle	Idle	Idle	Compact 2
21	Idle					
22						
23	THIRD		CYCLE			Idle
24	CYCLE					
25						
Operating time/cycle	8	8	4½	4½	6	9
Idle time/cycle	3	3	6½	6½	5	2

Figure 3.4   Multiple activity chart for concrete placing—original method

batching plant, the truck, two skips labelled 1 and 2, the crane, and the placing gang. A time scale is drawn vertically downwards and in each column a note is made as to whether each component of the system is working or idle. In this example the concrete batch size is 2 m³, the capacity of the truck is 4 m³ and the capacity of each skip is 2 m³. The method being employed is that a batch of concrete is mixed and placed into the truck which then waits for a second batch before travelling to the point where it fills the skips, within reach of the crane. The truck fills the first skip which is immediately hoisted by the crane and the truck then fills the second skip before it returns to the batching plant. After the first skip is filled it is hoisted into position, emptied and returned to ground level where the crane hook is moved from the first skip onto the second skip which then passes through the same cycle of operations. Reference to the final column in figure 3.4 will show, however, that the emptying of the second skip is delayed by the fact that adequate time has to be given for the compaction of the concrete by the placing gang. The whole sequence of operations in the placing cycle can therefore be set out in the multiple activity chart and the interrelationship between the individual components of the system can clearly be seen.

Examination of the chart will show that each component has a substantial amount of idle time and an attempt is now made to eliminate some of this. The overall cycle time is fixed at 11 minutes in this example by the truck which is working for 8 minutes each cycle, and then has an enforced wait of 3 minutes between receiving its two batches of concrete. If it were possible to eliminate the idle time of the truck an attempt could be made to reduce the overall cycle time to 8 minutes, but this is less than the time required by the compacting gang to place the concrete. It is likely however that it would be possible to increase the rate of compaction slightly, perhaps by the addition of another man, and hence it is assumed that this too could be reduced to 8 minutes. The key to this problem is to introduce a second truck, and of course both trucks could be of smaller capacity since they now need each only carry two cubic metres of concrete. The introduction of a second truck and the addition of an extra man to the placing gang could give rise to the revised method set out in the multiple activity chart of figure 3.5.

Examination of figure 3.5 suggests that it would be possible to use only one skip, since each is working for less than half the cycle time if the operation of changing the hook is eliminated. However, in practice the variability of hoisting, lowering and compacting time makes it desirable to introduce some flexibility into the system, and this can most cheaply be done by the use of the second skip. One further suggestion might be to cut out some of the double-handling of concrete by discharging directly from the batching plant into the concrete skip which itself was placed on the back of a truck. This would eliminate the need to discharge concrete from the truck into the skip at the crane pick-up point, but would of course necessitate the use of at least one additional concrete skip.

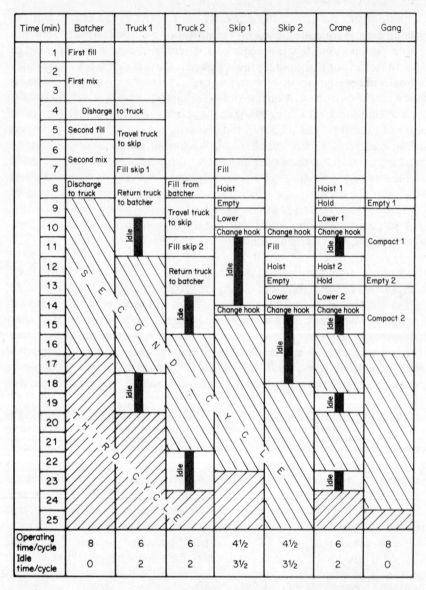

Figure 3.5　Multiple activity chart for concrete placing—revised method

Examination of the multiple activity chart, however, will show that the cycle time is now determined by the concrete-batching plant, and there would therefore be no point in reducing the time on other operations since they would then be subject to delay. In this particular case there is the additional problem that if concrete were placed in a skip and it travelled a considerable distance on the back of a lorry the vibration would be likely to compact the concrete sufficiently to prevent it from being easily discharged from the bottom-opening skip. This simple point illustrates the need for practical experiments to be undertaken before any revised procedure is finally adopted.

In this particular example it has been possible to reduce the cycle time by about 27 per cent, and since the objective was to minimise the amount of crane time being used this may be regarded as satisfactory. It would, however, in other circumstances be valuable to make an assessment of the cost difference between the original and the revised procedures. An assessment would also have to be made of the cost and technical problems involved in the alternative means of placing, whereby the concrete would be pumped directly from the batching plant into the shutters. Comparing here only the original and revised placing methods by crane the cost figures could be set out as in table 3.4.

Table 3.4   Cost comparison of original and revised concrete placing procedures

Unit	Original cost per hour	Revised cost per hour
Batcher	£12	£12
Truck 1	£4	£3·50
Truck 2	—	£3·50
Skip 1	£0·50	£0·50
Skip 2	£0·50	£0·50
Crane	£15	£15
Gang	£10	£12
Total	£42·00	£47·00
Cycle time	11 minutes	8 minutes
Cost per cycle	£7·70	£6·25
Reduction in unit cost is 19 per cent		

It was said at the beginning of this chapter that one of the features that typifies the construction industry is that projects are unique and are seldom, if ever, repeated in their entirety. It was also stated that many of the individual operations in a construction project are repeated and the above example of concrete placing is one such case. There is no doubt that in applications such as this the use of the multiple activity chart or any of the other work-study techniques is well suited. There are many useful books which explain the whole range of work study techniques available and give examples of their application.[1]

Time–Cost Interactions

Earlier in this chapter it was stated that in many construction situations time and cost will interact. This well-accepted fact usually becomes evident in contracts that run behind programme and thereby incur additional expenditure. One of the unfortunate consequences of delay is usually a corresponding increase in cost. An alternative view of the concept, however, is that if work is planned and executed well there may be the opportunity to bring about a reduction in project time by additional expenditure on the project. There may be many cases where additional expenditure on a particular operation is justified if it brings about a reduction in the overall project duration. In the terminology of network analysis, if the project duration can be reduced by cutting the length of the critical path, a comparison should be made of the cost and benefit of such a reduction. In many projects there will be a series of critical activities that offer the opportunity of a reduction in time brought about by the expenditure of additional sums in added plant or labour. There are very many examples of this situation but the following project gives a good illustration.

The Construction of a Road Bridge over a Railway

The project concerns the demolition of a masonry arch bridge over a main-line railway and its replacement with a new bridge of pre-stressed concrete construction. The nature of such a project is that there is much work to do in the diversion of services, removal of the existing bridge arch, reconstruction of the abutments, and other work prior to the installation of the new deck. There follows next the single operation of installation of the pre-cast pre-stressed concrete deck beams, which then permits the remainder of the work to be done including restoration of services, completion of the deck, construction of parapet walls, road surfacing and other finishes. The key to rapid completion of the project depends upon being able to install the pre-stressed deck beams quickly.

Since the bridge is over a main-line railway, access to the track for overhead working is only given between midnight on a Saturday night and 5 a.m. on a Sunday morning, that is, only five hours per week. Part of the terms of the contract specify that the pre-stressed concrete beams are supplied by the employer and delivered to the site on rail trucks. The contract documents include a suggested construction method that permits access on to one of the rail tracks for a crawler crane, provided the tracks themselves are protected by a sleeper mat. The estimated schedule for the installation of the pre-stressed beams is then as given in table 3.5, and as illustrated in figure 3.6.

Table 3.5 Original method for installation of bridge deck beams

Possession time allowed	5 hours
Installation time for sleeper mat	1 hour
Removal time for sleeper mat	1 hour
∴Available working time	3 hours
Time to install one beam	10 minutes
Number of beams installed per possession	18

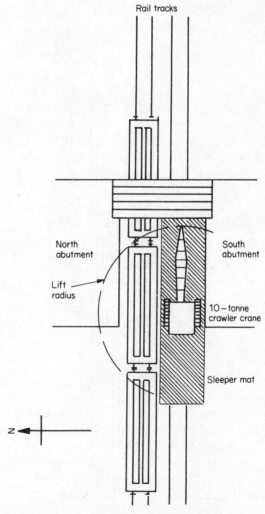

Figure 3.6 Installation of pre-stressed concrete beams—original method

There are 45 beams in total, and therefore three shifts will be required, but since only one possession is available per week the operation will take two weeks to complete.

This schedule makes use of equipment that is probably already on site, namely a 10-tonne crane with a 10-metre jib, but it is obvious that there is a lot of valuable possession time wasted. An alternative method of completing the work would be the 'sledge hammer and walnut' approach of using much larger equipment. This alternative is illustrated in figure 3.7 which shows the

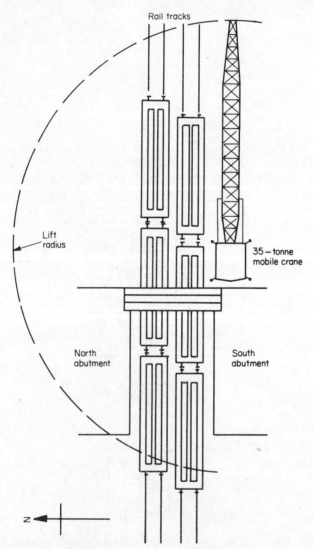

Figure 3.7   Installation of pre-stressed concrete beams—revised method

installation of a large 35-tonne crane with 25-metre jib erected completely outside the line of the railway track but with sufficient reach and capacity to lift beams from either of the rail tracks and install them in position on the abutments. This crane can be set up during normal working hours and can therefore make full use of the five-hour possession time for the purpose of installing beams. The fact that it has much greater reach and power also enables the time to install a single beam to be reduced from 10 minutes to 6 minutes and this can be seen in table 3.6. These factors together result in the whole operation being completed in one possession time.

Table 3.6 Revised method for installation of bridge deck beams

Possession time allowed	5 hours
Available working time	5 hours
Time to install one beam	6 minutes
Number of beams installed per possession could be 50	

There are only 45 beams to be placed, and therefore the whole operation can be completed in one overnight possession. This effectively gives a zero duration to the operation, since it is entirely outside normal working hours.

There is no doubt that the cost of the larger crane will considerably exceed the cost of the smaller crane, but in the particular circumstances of the project the extra expenditure may well be considered to be justified. This particular example may appear to be a special one, since the reduction in time worked from 15 hours to 5 hours results in an overall project saving of two weeks, but the situation is not necessarily a rare one. In many cases it may be preferable to concentrate additional effort on one or two critical activities in order to reduce the project duration, rather than to attempt to accelerate a project by an overall policy of overtime working. The latter may be vastly more expensive and unless control over the project is very carefully maintained the increased tempo of working at a higher rate over the whole project may result in confusion, which will more than offset the saving being sought by the acceleration. The broad principle to follow is to seek from the project programme the point where additional effort can most effectively be expended in an attempt to reduce the overall project time.

References

1. R. E. Calvert, *Introduction to Building Management*, Newnes-Butterworths, London, 1964.
2. Outline of Work Study, Parts 1 and 2, British Institute of Management, London, 1955/6.
3. R. Geary, *Work Study applied to Building*, Godwin, London, 1970.

4 Certainty, Uncertainty, Probability, Variability and Risk

In 1966 the Tavistock Institute published a study[1] on the building industry entitled 'Interdependence and Uncertainty', a title that evolved during the course of the research study to which it relates. The title really reflects two of the major findings of the research team, features that are well known to those involved in the construction industry, but not well understood. In chapter 3 reference was made to the fragmentation of the industry, and the interdependence between the fragments does impose a pattern of working upon the industry. The second feature, that of uncertainty, may well result from the complex interaction between the large number of organisations involved in the industry, but some aspects of uncertainty should be avoidable with good management and control. If careful work planning is undertaken, many of the doubts about delivery and completion dates, labour and material availability, quality of work, and so on can probably be eliminated; but it is of interest to note that the Tavistock research team formed the opinion that many people in the construction industry actually thrived on the uncertainty, and would feel at a loss if it were eliminated. It is unlikely that it will disappear, however, and even if it were possible to eliminate the difficulties that arise because other people do not work to programme, there will always remain some areas of uncertainty. These relate to problems of *chance* events in the future, for example the occurrence of floods, high tides, hurricane winds, all natural phenomena that cannot be accurately forecast. Other chance events concern the winning of contracts, where one company does not know whether it has bid above or below its competitors. The uncertainty in these cases arises from ignorance of the outcome of future events. Ignorance can also lead to uncertainty in relation to factual data; for example, information on ground conditions may be inadequate for a full understanding of the problems that will be involved in an earthworks contract. Here the word *ignorance* is used not in the common pejorative sense, but in the stricter sense of lack of information. This lack of information characterises many business and engineering decisions; the manufacturer of a new piece of building equipment may be ignorant of the market he can expect for his product; the civil engineering contractor may be ignorant of the exact level of the ground water on the site where he is about to commence work.

In both of these examples and many others there will be a *cost of ignorance*, that is, the costs that may arise by virtue of not having the right information at the time of making a decision. Similarly there is a *cost of information*, which is the cost incurred in seeking the required information, either by market survey in the case of the plant manufacturer, or the cost of drilling or other site investigation for the contractor. If the cost of ignorance is greater than the cost of information in a particular situation, then there is not doubt that the information should be sought and paid for.

In the preceding paragraphs two slightly different meanings have been given to the word *uncertainty*, first the shortcomings of systems that fail to function satisfactorily, and second the uncertainty associated with chance events in the future and the inherent ignorance of specific situations. It is the second meaning that is explored further here and later in chapters 8, 9 and 10; and it is important to show the distinction between certainty and uncertainty. One way to illustrate the distinction is to consider the problem of the design of a water tower as follows.

A water tower is to be built on top of a hill on an exposed site. The capacity of the tank and the level of water in the tank have been pre-determined and the design problem consists of a detailed design of the structure. The load due to the water in the tank is known almost exactly, and the self-weight of the tank can be determined within close limits, dependent only upon variation in density and dimension of the structure of the tank. The strength and elasticity of the materials used for construction, namely steel and concrete, are known reasonably closely even though some variation in concrete strength is likely. The stresses arising at various parts of the tank's structure may be a little more difficult to assess because of redundancies of the structural frame and the problems of stress distribution caused by the continuity of the structure. It may therefore be said that all of these factors referred to are known with certainty, that is, they may be determined with reasonable accuracy and moreover there is no doubt that such values as are determined will apply to the structure.

One major uncertainty that will apply to an elevated water-tower is the incidence of wind loading. We can be reasonably sure that during the life of a water tower it will definitely be subjected to winds of 100 km/h, but will definitely not be subject to winds of 300 km/h. Somewhere between these two limits are wind speeds that may or may not arise dependent upon some element of chance. The question arises as to whether the tank should be designed to withstand a wind speed of 150 km/h which is quite likely, 170 km/h which although rare may well occur, or a speed of say 200 km/h which has hitherto not been recorded in the area. It is reasonable to predict what the effects of any given wind speed would be and this calculation can form part of the design analysis for the tower.

In practice, design engineers overcome or side-step this problem by simply referring to standards that are laid down for the design of such structures. This is a perfectly reasonable thing to do since these standards are arrived at as a result of very careful detailed analysis of the probability of various wind speeds arising in different locations. There are, however, many areas of uncertainty that arise in the construction industry where no standards are available and where it is necessary for engineers or managers to make some calculation of the chances involved. It is in these cases that use can be made of various techniques that are frequently referred to under the heading of decision theory. It is often said that the major role of management is that of decision-making. While it is undoubtedly true that decision-making is the responsibility of the manager there are very many other functions that have to be fulfilled.

First, before any decisions can be made it is necessary to set up the relevant information and to show its reliability. Much successful management has been attributed to intuition, but conversely many failures can also be attributed to a lack of information. While it is certain that the role of management cannot be replaced by the techniques of management science, it is at the same time true that such techniques can be of considerable value to managers in helping them to arrive at rational decisions. Decision theory falls into this category since it is a technique or series of techniques that can help in tackling problems that were hitherto thought to be subject only to intuition and luck.

In the construction industry there is often a conflict in management between theoretically trained engineers and entrepreneurs who have 'come up the hard way'. The entrepreneur traditionally makes his decisions on the basis of intuition or hunch while the engineer searches for the deterministic solution to the problem that gives him one right answer. In decision theory we make use of the intuition of the experienced manager and combine it with the analytical ability of the engineer to produce a basis for taking decisions in problems where conditions of uncertainty apply. The examples given in chapters 8, 9 and 10 illustrate a range of applications of decision theory.

Certainty

This does not necessarily mean a precise knowledge of all the factors relevant to a situation, but it does mean that they are sufficiently well known for the purposes in hand. Many engineering design problems are concerned with certainty.

Uncertainty

This means that in the problem under consideration there is either a great degree of technical ignorance of relevant data, or there is a chance factor that may or may not arise, or the parameters concerned may vary over a wide range.

Probability

While many people throw up their hands in horror at the thought of uncertainty, the management scientist attempts to quantify uncertainty, and he does this through the concept of *probability*. This is not simply a descriptive term which says whether an event is highly probable or improbable, but is a measure of the real likelihood of its occurrence. The common illustration is to quote the example of very simple chance events such as the tossing of a coin, or the drawing of a playing card. If a coin is tossed a large number of times it can be expected to fall head up on half the occasions and tail up on the other half. There is thus a 50/50 chance of any one toss producing 'heads', and this is expressed as saying that there is a 50 per cent probability of a head, usually expressed $p_h = 0.5$. Similarly for a tail $p_t = 0.5$, and here it is noted that $p_h + p_t = 1$, or in words, the sum of the individual probabilities of all alternative outcomes of a chance event must be unity. If a card is drawn from a pack of playing cards there are clearly 4 chances in 52 that it will be an ace, that is $p_a = 4/52 = 0.077$. These are clearly chance events where the probabilities can easily be calculated from a knowledge of all possible outcomes that may arise.

Often in design or management problems such precise calculation of probability is not possible, but values can be given on the basis of past records, or simply by subjective judgement. Meteorological records will give data on wind speeds; for example, in the case of the elevated water-tower quoted at the beginning of this chapter we may say that there is a probability of 0.4 that wind in excess of 150 km/h will arise at some time during the next forty years. Since all possible outcomes must have probabilities that sum to 1.0, we may then conclude that there is a probability of 0.6 that there will be no wind in excess of 150 km/h within the next forty years. We may wish to be more specific than this and give a more detailed analysis of the probabilities of different wind speeds arising as follows.

No wind in excess of 125 km/h	Probability $p = 0.10$
Maximum wind in the range 125–150 km/h	$p = 0.50$
Maximum wind between 150–175 km/h	$p = 0.30$
Maximum wind between 175–200 km/h	$p = 0.09$
Maximum wind in excess of 200 km/h	$p = 0.01$

Note that the sum of the probabilities is 1.

Combination of Probabilities

It is sometimes necessary to consider the combination of a number of separate events that have distinct probabilities, and different forms of combination may apply. Returning to the case of tossing a coin, what is the probability of two heads in two throws? There are four possible outcomes of two throws, each with equal probability

$$\text{head} + \text{head} \quad p_{hh} = 0 \cdot 25$$
$$\text{head} + \text{tail} \quad p_{ht} = 0 \cdot 25$$
$$\text{tail} + \text{head} \quad p_{th} = 0 \cdot 25$$
$$\text{tail} + \text{tail} \quad p_{tt} = 0 \cdot 25$$

The probability of two heads is found by

$$p_{hh} = p_h \times p_h = 0 \cdot 5 \times 0 \cdot 5 = 0 \cdot 25$$

The probability of two tails is found by

$$p_{tt} = p_t \times p_t = 0 \cdot 5 \times 0 \cdot 5 = 0 \cdot 25$$

But the probability of one head and one tail is

$$p_{th} + p_{ht} = (p_t \times p_h) + (p_h \times p_t) = 2(0 \cdot 5 \times 0 \cdot 5) = 0 \cdot 50$$

Another simple game of chance is throwing dice, where on a single throw the chance of a 'four' is one in six throws, that is $p_4 = 0 \cdot 167$. The probability of two fours in two throws is $p_4 \times p_4 = 0 \cdot 167 \times 0 \cdot 167 = 0 \cdot 027$ and the probability of two similar throws of any number is

$$p_1 \times p_1 + p_2 \times p_2 + p_3 \times p_3 \ldots + p_6 \times p_6 = 6 \times 0 \cdot 027 = 0 \cdot 167$$

This last figure could clearly be predicted since if the first throw has produced any one number, then there is a one in six chance that the second throw will produce the same number.

Putting this concept into the context of the construction industry, the case of submitting tenders for contracts may be considered. A contractor X intends to submit a tender of £200 000 for a contract, and judges that at this figure there is a 90 per cent chance that he will be lower than competitor A, but only a 50 per cent chance that he will beat competitor B. Assuming that there are no other competitors his probability of winning the contract $p_X = p_A \times p_B = 0 \cdot 90 \times 0 \cdot 50 = 0 \cdot 45$. If there are more competitors in the running for the contract then the probability of a win for X is the product of the probabilities of beating each competitor individually, namely $p_X = p_A \times p_B \times p_C \ldots \times p_N$. This topic is considered further in bidding strategy, discussed in chapter 10.

In the area of construction planning it is valid to combine probabilities in order to find the probability of completion of a project by its scheduled date. A particular project consists of three separate areas of work R, S, T which run concurrently and independently, and for which the completion probabilities in table 4.1 apply. The probability of completion of all three sections of work by a specific month is given by $p_{RST} = p_R \times p_S \times p_T$. Note that in the example in table 4.1 the chance of completing by month twelve is little better than one in four, despite a high probability that area R will be complete, and in the worst area T there is a 50/50 chance of completion, a situation that is often tolerated.

Table 4.1 Combination of probabilities of project completion

Work area	Probability of this area being completed by month number				
	10	11	12	13	14
R	0·2	0·7	0·9	1·0	1·0
S	0·1	0·3	0·6	0·9	1·0
T	0	0·1	0·5	0·6	0·9
R+S+T	0	0·0210	0·270	0·540	0·90

The example illustrated in table 4.1 is an extremely simple one, and most construction programmes are very much more complex. It can easily be appreciated that in a practical project many of the activities interact, and it becomes extremely difficult to work out the combined probability of meeting an overall project-completion date. If in a project each separate activity has a shortest possible duration and a longest possible duration, then it is fairly straightforward to calculate the shortest possible project time by using only the shortest activity times. Similarly the maximum project duration can be found by calculation using only the longest activity times. These two limits will be very wide, and what is usually of interest is the *most likely* project completion date. Each separate activity will have a probability distribution between the two extreme limits, and these can be combined to find the probability distribution of overall project duration. The method of calculation is complex, but can be based on the network analysis methods described in chapter 2. The calculation is too complex to be included in this book, but is well described by Moder and Phillips.[2]

Cumulative Probability

There are many occasions when it is necessary to consider cumulative probability, as in the design of a temporary works coffer-dam described in chapter 8. It has already been shown that the chance of throwing a particular number with dice is 0·167, and this may now be expressed as a cumulative probability as follows in table 4.2. Clearly whatever happens the lowest number that can arise is one, and therefore the probability of one or more is 1·0. In the second row there is only one chance in six that the number will be lower than two, and therefore the probability that it will be two or more is given by $1·000 - 0·167 = 0·0833$. Each row is obtained by successively reducing the previous cumulative figure for the individual probability of that row.

Table 4.2   Cumulative probability of dice throwing

Probability that the number will be 1 or more	$1 \cdot 0$
Probability that the number will be 2 or more	$1 \cdot 0 - 0 \cdot 167 = 0 \cdot 833$
Probability that the number will be 3 or more	$0 \cdot 833 - 0 \cdot 167 = 0 \cdot 667$
Probability that the number will be 4 or more	$0 \cdot 667 - 0 \cdot 167 = 0 \cdot 500$
Probability that the number will be 5 or more	$0 \cdot 500 - 0 \cdot 167 = 0 \cdot 333$
Probability that the number will be 6 or more	$0 \cdot 333 - 0 \cdot 167 = 0 \cdot 167$

Returning to the water-tower example it is now possible to express the wind velocity probability in cumulative form. During the life of the tower it is considered certain that there will be a wind speed greater than 100 km/h, and hence

$$\text{wind in excess of } 100 \text{ km/h} \qquad p = 1 \cdot 00$$
$$\text{wind in excess of } 125 \text{ km/h} \cdot \qquad p = 0 \cdot 90$$
$$\text{wind in excess of } 150 \text{ km/h} \qquad p = 0 \cdot 40$$
$$\text{wind in excess of } 175 \text{ km/h} \qquad p = 0 \cdot 10$$
$$\text{wind in excess of } 200 \text{ km/h} \qquad p = 0 \cdot 01$$

Note that in this form the p values do not sum to $1 \cdot 0$, since each value includes each of the subsequent values, for example a wind of 160 km/h exceeds 100, 125 and 150 but not 175 or 200 km/h.

Variability

Statistical variation and variance are specific terms with very precise meanings that cannot readily be explained without a full introduction to the subject of statistics. There is not space in this book to cover statistical analysis adequately, but there are many good books available, starting with the popular introduction by Moroney.[3] The purpose of introducing the word variability here is to emphasise that in many problems associated with management, there is not always a consistent answer to a particular problem. In chapter 8 the subject of decision-making is discussed, and it is pointed out there that for any given situation where a decision has to be made the view of the decision-maker will be important, and the decision taken may vary from one person to another, or from one organisation to another. Even in simple design problems this is true, for example, quite different designs for the structural frame of a building would be produced by a consulting engineer, a reinforcement supply company, a structural steel-work manufacturer and a timber merchant. While this example may be trivial, there are other decisions that are affected by the size of a company, for example, a large organisation may be prepared to undertake a high-risk project, but a small company will not do so.

A second aspect of variability is that even when all the inputs to a system *appear* to be constant, the output may vary. Anyone who has worked out weekly site costs will know that there is great variability in production rates and costs, due to factors that were either thought to be irrelevant, or were quite unpredictable. The important point to note is that this variability does occur, and furthermore it may be possible to measure it. As was mentioned in chapter 1 it is often useful to quote the answer to a problem not as an absolute figure, but within a range. This is particularly applicable in the calculation of estimates, where it is much better to state that the cost of an operation will lie within the range £150 to £170, rather than simply say that it will be 'about £160'. It is essential in many cases to recognise that this variability can arise, for example, in the appraisal of a possible investment. Comparing the first two possible investments in table 4.3 it can be seen that while both projects A and B are likely to make the same profit, there is a possibility of a loss in project B, but not in project A. This fact would be most important in coming to a decision.

Table 4.3   Comparison of projects with variability of cost

Project	Cost (£)	Revenue (£)	Profit range (£)
A	30 000	32 000 to 38 000	2000 to 8000 profit
B	30 000	28 000 to 42 000	2000 loss to 12 000 profit
C	30 000	25 000 to 55 000	5000 loss to 25 000 profit

Risk

The above example serves to introduce the concept of risk, which is a measure of the chance of unacceptable losses arising. In the example in table 4.3 project B has the risk of a loss of £2000. Attitude to risk will vary from one person to another, and from one organisation to another, and this is discussed more fully in chapter 8. There is no reason why anyone should prefer project B to A since it offers no greater reward to compensate the risk inherent in project B. However, project C shows a rather larger risk but at the same time the possibility of a much greater profit and might be thought by some companies to be a risk worth taking.

Risk can also be related to factors other than cost, for instance the risk associated with the completion of a project on time, or the risk of contractors being omitted from tender lists by architects. Irrespective of the nature of the risk, it is possible to make some sort of estimate of its value in quantitative terms. This can then serve as one basis for the comparison of alternative courses of action.

References

1. *Interdependence and Uncertainty, A Study of the Building Industry,* The Tavistock Institute, London, 1966.
2. J. J. Moder and C. R. Phillips, *Project Management with CPM and PERT,* Van Nostrand Reinhold, New York, 1970.
3. M. J. Moroney, *Facts from Figures,* Penguin, London, 1969.

Part Two Certainty and Optimisation

Management in industry is often faced with complicated problems of a quantitative nature, which require more than simple arithmetic in their solution. For a manufacturer there will be one 'best' location for his warehouse that will minimise his transport costs; for a plant operator there is one 'best' job allocation that will minimise his production costs; in transport there is one 'best' route pattern to minimise running costs. These problems are of a static deterministic nature, or more simply the information is all available at the start of the analysis and the factors involved are not subject to random chance. There may be *variations* of the factors involved, but usually the extent of this variation is known. Chapter 5 describes the operational research methods of linear programming, transportation and allocation models, in relation to both design and management. Although it may appear to present a contradiction in terminology a dynamic programming approach to the static problem of drainage design is also included here. Chapter 6 deals with simple graphical models associated with the location of plant items, and chapter 7 deals with capital investment decisions, including the use of discounted cash flow.

5 Decision-making under Conditions of Certainty

In the previous chapter an explanation was given of what is meant, in this context, by certainty and this was compared with uncertainty. This chapter is concerned with giving some examples of methods that may be used under conditions of certainty.

Many readers may feel that nothing is certain in the modern industrial world and in particular nothing in the construction industry is completely predictable. However, as was pointed out previously we are not concerned with problem areas where we are able exactly to predict the outcome, but with cases where we can forecast with some confidence what will happen. For example, in many design problems we can predict that the structure will be subjected to a certain range of load without being able to say exactly what the load will be. In contract management we can forecast the costs that will be incurred in the construction of the work, again without being sure of exactly what the costs will be. On the other hand where we are concerned with cases in which there is a large element of random chance we are completely uncertain about the outcome; for example, when bidding for a particular contract we do not know at all whether or not we will be awarded that contract.

Confining interest now to situations where we are concerned with taking decisions under conditions of certainty, there is a large group of problems to which there is not one single right answer but a whole range of possible answers. The methods outlined in this chapter seek to find ways of tackling these problems in such a way that we may produce the best answer in a particular situation from any one particular point of view. This we refer to as optimisation. It should be carefully noted that the 'best' solution to a problem may depend on criteria of judgement. Consider for example the operation of a concrete-batching plant. It is usually possible to vary the proportions of coarse aggregate, sand, cement and water for a particular mix design within the strength constraint imposed by the design of the structure. This may be done to achieve certain objectives, but it may be difficult to decide which is the 'best' mix. There may be different criteria of assessment of the 'best' mix proportions as follows.

(1) The lowest material cost per cubic metre of concrete produced.
(2) The maximum daily output of the batching plant.

(3) The production of concrete that offers greatest ease of transport, placing and compaction with consequent lowest handling and placing costs.

One of the most common techniques used in the solution of problems of this type is known as linear programming. Explanation of this term is necessary before the method is described. In this context the word 'programming' does not mean setting out a plan of action, as in the case of contract planning or programming, but simply means the arithmetical manipulation of numbers, in rather the same way that we talk about computer programming. The word linear simply indicates that the relationship between the factors that we are considering can be represented on a graph by a straight line, and not by some rather more complicated form. Unfortunately the example of concrete-mixing plant operation previously quoted involves some factors that are not linear and therefore it does not offer a simple exemplification of the method. However, the following problem concerned with the operation of quarry plant associated with concrete-mixing plant gives an example where all the inter-relationships are linear.

Operation of Quarry and Concrete-batching Plant

For the purposes of this example it is considered that there are three main items of plant in the operation of the unit. The first of these is the washing plant which cleans the aggregate and frees it from unsuitable materials, the second is the screen which separates the aggregate into its separate particle-size categories, and thirdly the concrete-batching plant. Aggregate material may be bought from two different sources of supply, A and B. It is found that the qualities of these two aggregates differ in such a way that for the two materials different throughput rates on the washing, cleaning and mixing plants are achieved as given in the following table. The price of the two aggregates also differs but the ultimate selling price of the concrete is the same and it is assumed that the cost of cement is the same in both cases.

Table 5.1 Concrete plant production times and costs

		Aggregate A	Aggregate B
	Hours per week that each plant can work	Hours required per 100 m³ of concrete produced	
Washing plant	60	12	5
Screening plant	80	5	16
Batching plant	42	6	7
Cost and prices of concrete (£ per 100 m³)	Cement	1·50	1·50
	Aggregate	5·00	4·00
		6·50	5·50
	Gross profit	1·00	2·00
	Selling price	7·50	7·50

Note that the output rates for the washing and screening plant vary very considerably between the two aggregates A and B, probably more than they ever could in practice. These figures have been chosen in order to show more clearly the method being used; in practice their values might be rather closer but still significantly different. Each plant item is limited to a number of hours per week that it can work, this figure being a maximum that must not be exceeded, rather than an actual figure that will be achieved. These figures impose constraints upon the output from the plant as a whole, a feature that typifies problems that lend themselves to a linear programming application. Any combination of quantities of materials A and B put through the plant that keep within these constraints will offer a *feasible* solution to the problem, but there is likely to be only one optimal solution that will maximise the gross profit produced by the whole plant. There will be another, perhaps different, solution that will optimise the maximum *output*, in cubic metres, of the whole plant.

It may appear on examination of the cost figures in the table that since material B offers a gross profit per unit output of £2·00 compared with only £1·00 for material A, it is obvious that the plant should use only material B. However material B makes a very heavy demand on the screening plant which has a relatively low output rate when working on that aggregate. This means that the washing plant and batching plant are greatly under-utilised, and it may be possible to make better use of the plant as a whole by using some combination of aggregates A and B. It is possible to write a series of expressions that represent the constraints of the system and the desired objectives (per hundred cubic metres of concrete equivalent)

$$x_A = \text{volume of material A per week}$$

$$x_B = \text{volume of material B per week}$$

Washing plant

$$12x_A + 5x_B \leqslant 60$$

Screening plant

$$5x_A + 16x_B \leqslant 80$$

Batching plant

$$6x_A + 7x_B \leqslant 42$$

Gross profit $= 100x_A + 200x_B$

Volume output $= 100x_A + 100x_B$

The first expression says that for every x_A m^3 of concrete produced using aggregate A the washing plant will have to operate $12x_A$ hours, and for every x_B m^3 of concrete produced using material B the washing plant will have to operate $5x_B$ hours. It is known that the number of hours per week available on the washing plant is limited to 60 and hence we have the derivation of the expression that $12x_A + 5x_B$ must be equal to or less than 60. Similarly the expressions for the screening plant and batching plant are derived.

The next expression relating to gross profit is known as the *objective function*. It is known that each 100 m³ of concrete produced using aggregate A yields a profit of £100 and using aggregate B a profit of £200. It is then obvious that the gross weekly profit of the whole plant will be £$100x_A$ + $200x_B$. If we first seek to maximise the gross profit of the whole plant then it is this objective function or gross profit that we seek to maximise. An alternative objective function would be to maximise the volume of concrete produced per week and clearly this is simply $100x_A + 100x_B$. It is now possible to represent all these expressions in graphical form as shown in figure 5.1.

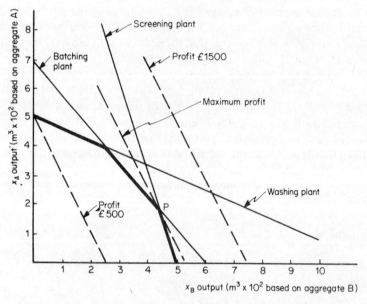

Figure 5.1   Concrete plant—constraints and profit functions

In the figure the line marked 'washing plant' is the line that represents the *equation* $12x_A + 5x_B = 60$. However the expression derived for the washing plant was based on the fact that the number of hours per week worked should not exceed 60, but could conceivably be less than 60. In the figure this means that any combination of x_A and x_B that lies on the line shown, or below and to the left of it, represents a feasible combination, *as far as the washing plant is concerned.* Similarly lines are drawn in the figure for the *equations* representing the screening plant and batching plant. Any solution that satisfies all three constraints must therefore lie to the left of and below all three lines in the figure. Such solutions must therefore lie within the area bounded by the heavy line. Note that negative values of either x_A or x_B are ignored.

It is possible to draw a line in the figure to represent the gross profit of the plant. For example if it were sought to achieve a profit of £1500 per week this could theoretically be done by using any combination of aggregate sources A and B that satisfied the expression $100x_A + 200x_B = 1500$, which is represented by the dashed line in figure 5.1. Unfortunately this line throughout its length falls outside the area of possible solutions bounded by the heavy line, and it is therefore concluded that it is not possible to achieve a profit of £1500 per week. If however, it were desired to achieve a profit of only £500 per week, also indicated by a dashed line in figure 5.1, it can be seen that any point on this line offers a feasible solution within the bounded areas and hence any combination of x_A and x_B on this line is possible and will achieve a profit of £500 per week. The profit lines for £500 and £1500 are parallel and any intermediate profit figure will be represented by another parallel line between these two. As the profit line is moved gradually to the right, parts of it start to fall outside the area of possible solutions bounded by the heavy line, and eventually a position is reached where only one point of the profit line offers a feasible solution, that is, where it is coincident with one of the extremities of the bounded area at point P in figure 5.1. At this point the solution has been optimised by maximising the objective function for gross profit. The values for x_A and x_B, the respective quantities of aggregates A and B corresponding to this solution, can either be read from the graph or calculated by solving as simultaneous equations the expressions representing maximum availability of the batching plant and screening plant as follows.

$$5x_A + 16x_B = 80 \qquad (7.1)$$

$$6x_A + 7x_B = 42 \qquad (7.2)$$

$$60x_A + 96x_B = 480 \qquad (7.1) \times 6$$

$$60x_A + 35x_B = 210 \qquad (7.2) \times 5$$

subtracting

$$61x_B = 270$$

therefore

$$x_B = 4 \cdot 43$$

and substituting gives

$$x_A = 1 \cdot 82$$

The gross profit would be

$$100 \times 1 \cdot 82 + 200 \times 4 \cdot 43 = £1068$$

and the output would be

$$100 \times 1 \cdot 82 + 100 \times 4 \cdot 43 = 625 \text{ m}^3$$

Turning to the alternative objective of maximising the output from the whole plant it is possible to repeat the procedure by representing the objective function of volume in the same graphical form. This is indicated in figure 5.2 where a series of parallel lines is shown for different volumes of output. In this case the objective function is maximised at point Q at which the following values apply

$$x_A = 3 \cdot 88$$

$$x_B = 2 \cdot 67$$

$$\text{gross profit} = £922$$

$$\text{output} = 655 \text{ m}^3$$

These calculations show that the point of maximum profit is not necessarily coincident with the point of maximum output, but of course with different values for the constraints they could indeed be coincident.

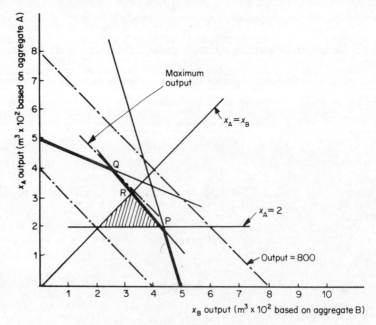

Figure 5.2 Concrete plant—constraints and output functions

It is a simple matter to impose the effect of other constraints upon production of this type of plant. For example it could be that there is a contractual obligation to produce at least $200 \, \text{m}^3$ per week of concrete using aggregate A, and this constraint would be shown graphically as in figure 5.2 by the line represented by the equation $x_A = 2$. This would restrict the range of possible solutions to those lying within the bounded area and above the line $x_A = 2$. Another constraint might be that for other reasons at least 50 per cent of the concrete used should be based on material B. This would be represented graphically by the line $x_A = x_B$, with possible solutions lying all to the right of the line, that is, where x_B is greater than x_A. This further restricts the area of possible solutions which must now lie within the shaded area on figure 5.2. The objective function for gross profit maximised at point P is nearly a feasible solution since it lies just outside the shaded area, but the solution at point Q which maximises the output is no longer a possible solution since it arises at a point where the value of x_A exceeds the value of x_B. The objective function for output would now be maximised at point R at which the values are as follows

$$x_A = 3 \cdot 23$$

$$x_B = 3 \cdot 23$$

$$\text{gross profit} = \pounds 969$$

$$\text{output} = 646 \, \text{m}^3$$

It has been possible to solve this problem graphically because there were only two materials A and B being considered. If a third material C were introduced with differing requirements on the various plant items, solution of the problem would become a little more complicated. It would be possible to think in terms of a three-dimensional graphical solution using the two axes for x_A and x_B as before together with a third axis x_C perpendicular to the page. The constraints would no longer be represented by lines on the diagram but by plane surfaces in the three-dimensional graph, and the range of possible solutions would be represented by a solid bounded by a number of such planes. The objective function of gross profit could be represented by a series of parallel planes and the maximum would be found by selecting that plane which just touched one of the corners of the solid. While this approach may be cumbersome it is feasible, but if four or more variables are involved the graphical approach becomes impossible.

In general terms, however, the method of seeking to maximise an objective function subject to a large number of constraints offers a solution to the linear-programming problem, but the evaluation of the answer has to be undertaken mathematically rather than graphically. One of the most common techniques is the *simplex method* which requires for its solution an understanding of matrix algebra. It is considered that the method is outside the scope of this book, but for those interested in the application of the method a number of appropriate references are given.[1,2,3,4,5] For those who wish to use the method but do not wish to undertake the evaluation themselves there are available a number of computer packages that will produce an answer. Indeed where the number of constraints and variables is large, it is probably essential to use a computer, since calculation by hand would be extremely time-consuming and, more importantly, liable to error. In this book therefore a number of applications are described without their solutions being given in detail.

Design of Structural Steel Frame

For most purposes the design of structural steel framework will seek to minimise cost consistent with satisfactory performance in relation to strength and deflection. The linear-programming method may be useful in the design of steel frames on the assumption that it is possible to derive a linear relationship between the cost and strength of steel sections. This may not always be true but within limits it is usually possible to derive such an expression. The application is based upon the principle of virtual work, which says that the work done by the load moving through a distance equal to the deflection of the load point is exactly equal to the work done internally within the structure by the yielding of its members. The failure of a steel frame may be said to occur when the loading reaches a level such that plastic hinges are formed at one or more points and the structure becomes a mechanism.

The analysis consists first of setting down all the possible modes of failure of the structure considering all possible combinations of plastic hinges necessary to create a mechanism. For each mode of failure it is then possible to write down an expression which relates, by the principle of virtual work, the following expression

$$\sum_1^n P\delta \leqslant \sum_1^m M\theta$$

where

> P_1, P_2, etc. are the loads
> δ_1, δ_2, etc. are the corresponding deflections of each load
> M_1, M_2, etc. are the ultimate moments of resistance at each plastic hinge
> θ_1, θ_2, etc. are the rotations of each plastic hinge

The load-point deflections are geometrically related to the yield rotations at the hinges and therefore each expression can be written purely in terms of load and moment of resistance. For any particular combination of loads it is then possible to write a series of expressions representing the modes of failure of the structure, and each of these will represent a combination of various M values where each M value represents the ultimate moment of resistance of one of the members of the frame. The objective function relates the M values to either the total weight or total cost of the structure, which ever it is we seek to minimise, hence for each loading combination it is possible to optimise the design of the structure, and by considering the various possible loading patterns the whole design may be optimised.

Design of a Slab Supported at Four Corners

A slab supported at four corners must be designed to withstand punching shear at the supports, bending failure, and excessive deflection. The design of the slab to withstand bending may be undertaken again using the principle of virtual work as in the previous example. As before the principle states that the work done by the load in passing through its failure deflection must equal the work done within the structure by the yielding of the slab. This case is a relatively simple one for a single rectangular slab supported at the corners, since there are only three modes of failure that need be considered, as shown in figure 5.3. The slab could fail either along its major axis, its minor axis or along both diagonals simultaneously. Under any of these modes of failure the centre of the slab will deflect an amount δ and if the slab carries a uniformly distributed load then the average distance deflected by the load

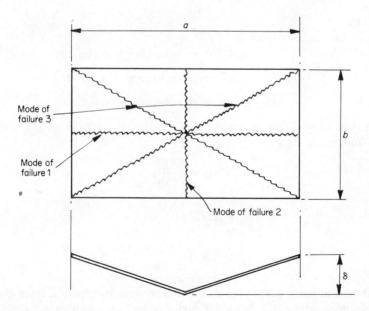

Figure 5.3   Modes of failure of rectangular slab supported at corners

will be $\delta/2$ in modes 1 and 2, and $\delta/3$ in mode 3. The relationships connecting load and moment of resistance are then as set out below

Mode 2
$$P\frac{\delta}{2} \leqslant 2\delta\frac{2}{a}M_b$$

Hence

$$P \leqslant \frac{8M_b}{a}$$

Mode 1
$$P\frac{\delta}{2} = 2\delta\frac{2}{b}M_a$$

Hence

$$P \leqslant \frac{8M_a}{b}$$

Mode 3
$$P\frac{\delta}{3} \leqslant 2\delta\frac{2}{a}M_b + 2\delta\frac{2}{b}M_a$$

Hence

$$P \leqslant \frac{12M_b}{a} + \frac{12m_a}{b}$$

where

$$P = \text{total load}$$
$$\delta = \text{deflection at the centre of the slab}$$
$$M_a = \text{moment of resistance along } a\text{-axis}$$
$$M_b = \text{moment of resistance along } b\text{-axis}$$

Consider now a specific case where $P = 1000$, $a = 20$, $b = 10$. The three inequalities now become

$$1000 \leqslant \frac{8M_b}{20}$$

$$1000 \leqslant \frac{8M_a}{10}$$

$$1000 \leqslant \frac{12M_b}{20} + \frac{12M_a}{10}$$

If it is assumed that the objective function seeks to minimise the moment of resistance (since it is assumed that this will also minimise the weight and/or cost) then the objective function is simply $M_a + M_b$. These may all be drawn graphically, as shown in figure 5.4. This example turns out to be almost trivial, since the result could have been predicted by 'common sense'. However, it demonstrates the approach in such a way that its validity is self-evident, namely that a minimum moment of resistance must be achieved along both major axes, and that if failure occurs it will be along one of them, not along both diagonals.

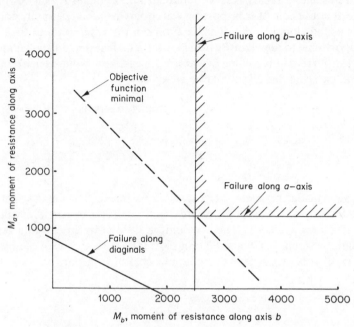

Figure 5.4   Slab design—constraints and strength

The Transportation Problem

This name is used to cover a group of problems which may be tackled by standard linear-programming methods, but which can also be tackled by a rather less mathematically involved technique. The technique has been derived, and is largely used, by distribution industries. It is typified by the problem situation where a company has goods in a number of depots around the country and has to distribute them to customers in various locations. In this case the constraints are that at each depot there is available no more than a certain amount, and an exact amount has to be delivered to each customer. The objective function in this case is to minimise the total transport costs. The delivery of materials such as concrete aggregates, cement and bricks from various sources to different sites are examples of this problem which may be tackled by the transportation method. The method also can be used in other cases such as the following one concerned with the design of a water-supply system.

Water-supply Design as a Transportation Problem

This example considers the lay-out design of a supply system where four reservoirs p, q, r, s, are available to supply water to three towns A, B, C. Figure 5.5 shows the locations of reservoirs and towns together with the distances between them by possible pipe routes. Also are shown the daily quantities available at each reservoir and the daily demand at each town. It is assumed that the cost of transporting water between the reservoirs and the towns is proportional to the distance between the reservoir and town and also proportional to the quantity transported. The first step is to calculate the cost of transferring unit volume of water from each reservoir to each demand point and entering these into a table as follows.

		Supply reservoir			
		p	q	r	s
Demand town	A	10	13	14	20
	B	7	8	29	23
	C	28	27	16	14

In this particular problem it is necessary to ensure that the cost entered into this table is the lowest appropriate cost in each case, since there will be a choice of routes available. There would be little point in carrying out an optimisation calculation when the unit costs for each route were themselves not optimal. It is fairly clear that the unit cost of transport of water from

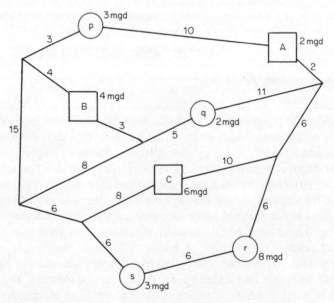

Figure 5.5   Water supply and demand locations

reservoir p to town A will be 10 and this figure is therefore entered into the table. Similarly the unit cost of transporting water from q to A will be $11+2=13$ which is therefore entered into the table. It is not difficult with the particular numbers used in this problem to fill in the table but if any readers find difficulty in selecting the minimum-cost route for a particular reservoir to town supply, then the method of least-cost route-planning described in chapter 6 in this book is appropriate.

As is often the case in transportation problems the supply does not equal the demand and the problem is therefore out of balance. In this case we have a surplus of supply and in order to keep the problem in balance it is necessary to create a fictitious demand point D with sufficient demand to take up the excess of supply over the demand from the other towns A, B and C. It is then assumed that the cost of transporting from each of the reservoirs p, q, r and s to town D is zero, this being equivalent to saying that there is no cost involved in transporting water to an imaginary town, in other words keeping it in the reservoir. The method of solution is by a series of tables as set out in figures 5.6, 5.7 and 5.8.

Across the top of the table the supply reservoirs p, q, r, s are set together with their available quantities in millions of gallons per day. Down the left-hand column of the table the towns are listed together with their daily demand in millions of gallons per day. Below and to the right of the heavy line there is now a matrix of the possible transport routes used. In the bottom right-hand corner of each box in this matrix are entered the transport costs

Supply / Demand		p 3	q 2	r 8	s 3
A	2	**1** 10	13	14	**1** 20
B	4	**2** 7	**2** 8	29	23
C	6	28	27	**4** 16	**2** 14
D	4	0	0	**4** 0	0

Figure 5.6 Calculation of optimal distribution pattern—original allocation

for each possible route, noting that all routes to town D have zero cost. The procedure now is to assume a possible solution to the problem and then by iteration to try to improve it.

Rather than simply make a random allocation of routes used it is sensible to use to the maximum routes with lowest costs. In this case all four routes to town D have zero cost and therefore it is necessary to decide from which reservoir town D should be supplied. The reservoir that has generally the highest supply cost is r and therefore it appears to make sense to use the route r to D. The number 4 is therefore entered to that box in the table thereby filling all of town D's requirements. There are however still four units left at reservoir r and it would be sensible to allocate these either to A or C where costs are relatively low compared with town B. In this case four units have been entered on the route r–C. This now takes the full supply available from reservoir r but does not yet satisfy town C, for which another two units are required to fulfil its demand of 6 million gallons per day. The cheapest way of obtaining these two units are from reservoir s, therefore the figure 2 is entered into the route s–C. This in turn leaves one unit of water available from s and the cheapest location to send this to is A. A now requires another one unit and the cheapest source of this is p. The figure 1 is therefore entered into the route p–A leaving two units available from p which are most cheaply transported to town B. This leaves town B still short of two units which can exactly be met from reservoir q. All demands have now been taken up and all supplies met, and a feasible solution to the problem has been attained. It is likely however that it is not the optimal solution to the problem, which is what is now sought. In a problem as simple as this one it is not difficult to make a sensible attempt at an initial solution, as described above. In other cases where a sensible initial solution is not so obvious many iterations of the solution may be needed. In these cases it is useful to use Vogel's approximation method[5] to find an initial solution.

The next step uses the concept of 'shadow costs'. This concept is that each route can be split into two elements—a dispatch charge from the supply end and a receiving charge from the customer end. Writing that p, q, r, s are respectively the supply costs from reservoirs p, q, r, s and A, B, C, D are respectively the receiving costs at A, B, C and D it is then possible to write an equation for each route cost for each route used as follows

$$p + A = 10 \qquad p = 10$$
$$p + B = 7 \qquad B = -3$$
$$q + B = 8 \qquad q = 11$$
$$r + C = 16 \qquad r = 22$$
$$r + D = 0 \qquad D = -22$$
$$s + A = 20 \qquad s = 20$$
$$s + C = 14 \qquad C = -6$$

Here there are seven equations for eight unknowns and it is only possible to obtain a solution of these by arbitrarily assigning a value to one of the variables. If A is given the value 0 it is then possible to work out the other shadow costs as shown on the right-hand side of the tabulation above.

If it can be shown that, by using one of the routes so far not used, there would be a resultant saving in total cost, then this implies that the optimum distribution pattern has not been found. If it can be shown that for any *unused* route, the route cost minus the sum of the shadow costs is negative then a saving would arise from using that route. The next step is set out in figure 5.7 which is a table similar to that in figure 5.6 with the addition of the

Shadow costs			10	11	22	20
			p	q	r	s
		Supply / Demand	3	2	8	3
0	A	2	**1** 10	(+2) 13	(−8) +1 14	1−1 20
−3	B	4	**2** 7	**2** 8	(+10) 29	(+6) 23
−6	C	6	(+24) 28	(+22) 27	4−1 16	2+1 14
−22	D	4	(+12) 0	(+11) 0	**4** 0	(+2) 0

Figure 5.7 Calculation of optimal distribution pattern—first iteration

shadow costs just worked out, that is, p, q, r, s and A, B, C, D. The route costs are entered in the bottom right-hand corner of each box as before. In the top left-hand corner of each box representing an *unused route* is entered in brackets a figure equal to the route cost minus the sum of shadow costs. Since one of these figures in brackets has a negative value it can be concluded that the optimal solution to the problem has not yet been found, and in order to achieve the optimal solution it will be necessary to make use of the route which shows a negative value here. The arbitrary solution previously used is entered into the boxes in the table. The route r–A shows a negative value for route cost minus the sum of the shadow costs, therefore an allocation is made to this route of an amount that will reduce one of the other routes, probably a more expensive one, to zero. Since s–A is the more expensive of the other allocations to town A the allocation there is reduced by one unit. This however leaves a surplus supply at reservoir s and therefore the allocation to one of the other towns is increased, in this case town C. This is turn means that town C is now being supplied with more than it needs, and it is necessary to make a reduction in supply from another reservoir, in this case r. A reduction of one unit allocation on the route r–C then balances the increase of route allocation r–A and the problem is again in balance. The solution arising from this modification must now be tested in the same way to see whether or not it is optimal.

The test consists of a further iteration of the problem as set out in figure 5.8. The route costs are entered in the boxes as before together with the route allocations deriving from the previous iteration. A new set of values for shadow costs is then calculated from the equations as follows, putting arbitrarily the value $A = 0$.

$$p + A = 10 \qquad p = 10$$
$$r + A = 14 \qquad r = 14$$
$$p + B = 7 \qquad B = -3$$
$$q + B = 8 \qquad q = 11$$
$$r + C = 16 \qquad C = 2$$
$$s + C = 14 \qquad s = 12$$
$$r + D = 0 \qquad D = -14$$

These new shadow costs are entered into figure 5.8 and again for the unused routes a figure is calculated for the route cost minus the sum of the shadow costs. This time it is found that in no case does this figure have a negative value, indicating that an optimal solution has been found. In this problem it is proved possible to achieve an optimal solution with only two iterations but it would commonly take many more than this to arrive at a solution. Having found the optimal allocation to routes it is a simple matter to calculate the sum cost by summing the products of allocation and route cost. This can be easily read from figure 5.8 and is equal to $1 \times 10 + 1 \times 14 + 2 \times 7 + 2 \times 8 + 3 \times 16 + 3 \times 14 + 4 \times 0 = 144$.

The final optimal solution for this particular problem can then be stated as follows.

Reservoir p should supply 1 million gal/day to town A and 2 million gal/day to town B

Reservoir q should supply 2 million gal/day to town B

Reservoir r should supply 1 million gal/day to town A and 3 million gal/day to town C

(and retain its balance of 4 million gal/day)

Reservoir s should supply 3 million gal/day to town C

Shadow costs			10	11	14	12
			p	q	r	s
		Supply / Demand	3	2	8	3
0	A	2	(+2) **1** 10	13	**1** 14	(+8) 20
-3	B	4	**2** 7	**2** 8	(+18) 29	(+14) 23
2	C	6	(+16) 28	(+14) 27	**3** 16	**3** 14
-14	D	4	(+4) 0	(+3) 0	**4** 0	(+2) 0

Figure 5.8 Calculation of optimal distribution pattern—second iteration

Degeneracy

In the solution of the transportation problem it sometimes occurs that there are two less routes in use than the number of shadow costs to be determined and it is therefore not possible to evaluate the shadow costs by solving the equations as was done above. This problem can be overcome by arbitrarily assigning a very small quantity ε to one of the unused routes, this simply being a device to enable shadow costs to be worked out without having any effect on the solution. A full description of the transportation method is given in the references.[1,2,3,4]

The Assignment Problem

The assignment problem is a particular case of the transportation problem in which there is only one item at each dispatch point and only one item required at each receiving point. This means that there must be equal numbers of dispatch and receiving points *and also routes*. Computation by the previously mentioned method would be highly degenerate, for example in a 4×4 matrix there would only be four routes used and it would be necessary to add small quantities to three other routes in order to solve the problem. This could be done but there is a simpler method of solving the assignment problem, as described in the following example.

Allocation of Mobile Cranes to Construction Sites

A crane-hire company operates from five depots p, q, r, s, t, spread throughout a region and at each depot it has based one large mobile crane in addition to other plant. On a particular day orders have been accepted to deliver these five cranes to five separate sites A, B, C, D, E, also spread throughout the region. The problem is to decide which crane should be sent to which site in order to minimise the total travelling distance.

The solution is set out in matrix form in table 5.2 showing the distance from each depot to each site. In a matrix it is possible to deduct any number from all the elements in a row (or column) without affecting the answer other than by reducing the total distance covered by the same number. The first step is to deduct the smallest element in each row from all other elements in the row as is done in table 5.2b. By a similar procedure the next step is to reduce the values in the matrix by deducting the smallest element in each column from every element in the column, as in table 5.2c. The next step is to try to make a 'zero assignment' by drawing lines through the elements of the matrix in such a way as to cross out all the zeros with the minimum possible number of lines. In table 5.2c it is only necessary to use three lines to cross the zeros and since this is less than the number of rows in the matrix, a zero assignment has not been achieved. The next step is to find the smallest uncovered element and deduct its value from every element in the matrix and then re-add it to all the elements covered by a line, which in this problem results in table 5.2d. Note that this means that uncovered elements in this matrix have been reduced by two units, elements covered by only one line remain as they were since the number two has been deducted and added back, but elements that were covered by two lines have been increased by two since they were first reduced by two units and subsequently two units added back twice.

Another attempt is made to draw lines through the matrix to make a zero assignment, this time resulting in four lines being necessary to cross out all the zeros. This means that a zero assignment has still not been achieved and it is necessary to repeat the process again, as shown in table 5.2f. This time it has been necessary to use five lines to cross out all the zeros and therefore a zero assignment has been achieved. The actual assignment is determined as follows. Column 5 has only one zero and therefore the route t–E must be used, and other assignments to E can be ignored. Similarly the route s–D must be used and other zero routes to D ignored. Row C has only one zero namely p–C meaning that p–B must be ignored. This means that route q–B must be used and finally also the route r–A. It is now possible to go back to the original matrix of route distances in table 5.2a and sum the total distance travelled namely r–A + q–B + p–C + s–D + t–A = 10 + 7 + 8 + 10 + 10 = 45 units.

Table 5.2a Matrix of route distances

	p	q	r	s	t
A	3	5	10	15	8
B	4	7	15	18	8
C	8	12	20	20	12
D	5	5	8	10	6
E	10	10	15	25	10

Table 5.2b Matrix with lowest value
deducted from each element in rows

	p	q	r	s	t
A	0	2	7	12	5
B	0	3	11	14	4
C	0	4	12	12	4
D	0	0	3	5	1
E	0	0	5	15	0

Table 5.2c Matrix with lowest value
deducted from each element on columns

	p	q	r	s	t
A	0	2	4	7	5
B	0	3	8	9	4
C	0	4	9	7	4
D	0	0	0	0	1
E	0	0	2	10	0

Table 5.2d Matrix with elements
reduced—first iteration

	p	q	r	s	t
A	0	0	2	5	3
B	0	1	6	7	2
C	0	2	7	5	2
D	2	0	0	0	0
E	2	0	2	10	0

Table 5.2e   Matrix with elements
reduced—second iteration

	p	q	r	s	t
A	1	0	2	5	3
B	0	0	5	6	1
C	0	0	6	4	1
D	3	0	0	0	0
E	3	0	2	10	0

Table 5.2f   Matrix with elements
reduced—third iteration

	p	q	r	s	t
A	1	0	[0]	3	3
B	0	[0]	3	4	1
C	[0]	1	4	2	1
D	5	2	0	[0]	2
E	3	0	0	8	[0]

Drainage Design Problem Tackled by Dynamic Programming

Dynamic programming and decision-tree methods (described in chapter 9) are typified by being able to cope with a sequence of decisions that will be affected by the outcome of decisions previously made relating to other factors. The use of the word 'previously' usually means 'coming earlier in time' but it can be applied to a flow system to mean further back in the system. This means that the method can be used in the solution of deterministic problems such as drainage design, where we are not concerned with the uncertainty associated with future events. Therefore the following example is included here rather than in chapter 9 where uncertainty is involved (it is not *necessary* to have read through chapter 9 before tackling this problem).

A simple exemplification of the dynamic programming method is in the following problem illustrated in figure 5.9. A main drain ABCD is to be constructed so as to collect water from three cross-feeders, A, B and C, and discharge them to a river at D. The maximum possible inflow at each of A, B and C is $0·3$ m³/s and therefore the main drain has to be designed to carry the following

> capacity of length AB must be not less than $0·3$ m³/s
> capacity of length BC must be not less than $0·6$ m³/s
> capacity of length CD must be not less than $0·9$ m³/s

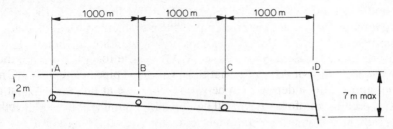

Figure 5.9   Longitudinal section of main drain

If it is assumed that the most economical arrangement is to design the pipe under full-flow conditions, it is well known that the flow through the pipe is some function of the diameter of the pipe and the slope to which it is laid. The actual relationship is given by the formula

$$Q = K(\phi)^{8/3}s^{1/2}$$

where Q is the quantity flowing in m^3/s
       K equals a constant (in this case taken to be 30)
       ϕ equals the diameter of the pipe in metres
       s equals the gradient of the pipe expressed as a fraction

The cost of the drain consists of two major elements, namely to supply, lay and joint the drain pipes, and excavation and backfill costs. The most economical design is clearly the one that minimises the sum of all these costs consistent with a capacity that is just but not more than adequate to take the required flows. If we consider first the length of drain AB it is clearly possible to select a number of combinations of diameter and gradient of pipe that will represent full-flow conditions at a flow of $0 \cdot 3$ m³/s. The diameter and slope are inversely related and therefore if we select a relatively small diameter of pipe (say 525 mm) we shall have to lay it at a relatively steep gradient. While this economises on purchase price of the pipe the steep gradient will mean that the drain will be at a considerable depth at point B and therefore the excavation costs will be relatively high. If conversely we select a large diameter of pipe it will be able to cope with the required flow at a flat gradient. In the second case the purchase price of the pipe will be greater but due to the lower gradient the excavation costs will be relatively low. It is a fairly simple matter to calculate the total cost of the drain length from A to B for various diameters of pipe and from these to select the cheapest design for that length.

This is, however, only part of the problem since the drain has to continue beyond B along the length BC with an increased capacity. The same criteria of diameter, slope and cost would apply, but the important point to note here is that the starting depth at point B has been *previously determined* by the decision made on the design of length AB. If we have selected a small-diameter pipe for AB with its consequent steep gradient we shall then have committed the length BC to start at a considerable depth. It might have been better to use a larger diameter pipe for AB even though it were not the cheapest for AB so that the length of pipe from B to C may commence at not too great a depth. The way to arrive at the cheapest design for the length ABC therefore, would be to consider all possible combinations of diameter and gradient and sum the component costs for each combination. Exactly the same argument holds for consideration of the final section CD of the pipe. The most economical design for the whole pipe therefore is to consider all possible combinations of diameter and gradient for the three lengths of pipe, bearing in mind that the depths are interactive, that is, the gradient chosen for AB will determine the depth of point B, and this together with the gradient of BC will determine the depth of point C, and so on. This is a somewhat formidable calculation to undertake if several different pipe diameters are considered.

In the case being described there are eight possible pipe sizes for each length, and there are therefore $8 \times 8 \times 8 = 512$ possible combinations to calculate. This task would be virtually impossible by hand, but could easily be undertaken by computer. This is a typical situation where a computer does not merely carry out calculations rapidly, but by virtue of its capacity to handle vast amounts of arithmetic it can be used in methods that would otherwise be extremely cumbersome. Here we are concerned however with a short-cut approach that does not require the use of a computer. The method is essentially the same as that used to solve the number–square problem by dynamic programming, in chapter 9.

The first step in tackling the problem is to calculate, using the formula given above, the fall over 1000 m necessary for each pipe size to carry each of the required flows. These are set out in table 5.3.

Table 5.3   Fall between feeders for various flows and pipe sizes

Pipe diameter(mm)	Fall (m) over 1000 m length for a flow		
	$Q = 0{\cdot}3$ (AB)	$Q = 0{\cdot}6$ (BC)	$Q = 0{\cdot}9$ (CD)
525	3·125	12·500	28·125
600	1·525	6·100	13·720
675	0·814	3·256	7·330
750	0·464	1·855	4·174
825	0·279	1·116	2·511
900	0·175	0·702	1·575
975	0·114	0·458	1·026
1050	0·077	0·308	0·693

The other necessary data is the cost of supplying, laying and jointing the pipe, and the excavation and backfill costs as are set out in table 5.4.

Table 5.4   Pipe purchase and laying costs

Pipe diameter (mm)	Supply lay and joint (£/m)	Excavation costs for various depths (£/m run)					
		up to 2 m	3 m	4 m	5 m	6 m	7 m
525	8·00	4·90	8·90	13·30	19·80	25·40	31·40
600	9·70	5·10	9·30	14·00	21·00	26·70	32·70
675	12·00	5·40	9·70	14·80	22·20	28·00	34·00
750	18·00	5·70	10·20	15·50	23·40	29·30	35·30
825	19·70	6·00	10·60	16·30	24·70	30·60	36·60
900	25·00	6·30	11·00	17·00	26·00	32·00	38·00
975	30·30	6·60	11·40	17·80	27·20	33·40	39·40
1050	35·00	6·90	11·90	18·70	28·50	34·90	40·90

The calculation procedure is set out in detail in table 5.5 with the following explanation. For the purposes of this example, incremental depths of 0·5 m have been considered, and the feasible depth increments along the pipe have been set out. For example the first row in table 5.5 shows a depth of 2·0 m at A, 2·5 m at B, 3·0 m at C. Bearing in mind that the maximum depth of D is 7·0 m the most economical pipe size for length CD, starting at a depth of 3·0 m at C, is determined by calculating the total cost from the data in tables 5.3 and 5.4., and is found to be 900 mm with a total cost of £45 400 which is entered in column 11 of table 5.5.

A similar calculation for the second row in table 5.5 shows that the most economical pipe size if C is at a depth of 3·5 m is 825 mm, with a cost of £45 600 for the length CD. Further calculations are made for all depths of C from 4·0 to 6·0 m in 0·5 m increments. Considering again the first row of table 5.5 the length BC is examined. The constraints here are that the pipe is limited to a fall from 2·5 m to 3·0 m and the most economical pipe size that can cope with the flow along BC with a fall not exceeding 0·5 m is the 975 mm size, at a cost of £41 700 for the length BC. The second row indicates the best pipe size, if the fall is restricted to 1·0 m (3·5 m − 2·5 m), to be 900 mm at a cost of £38 400. These cost figures are entered in column 7. The same procedure is followed for the length AB where in the first row the most economical pipe size is found to be 750 mm at a cost of £28 200, which is entered in column 3.

The table is set out in such a way that the cost figures for length CD are repeated down the table for each different value of the depth at B, noting that one value is omitted from each block moving down the table since the depth at C must always be greater than the depth at B. Using the dynamic programming approach described in chapter 9 this is effectively answering

Table 5.5　Calculation of cumulative costs for main drain

1 A is at (m depth)	2 Pipe size (mm)	3 Cost AB (×£1000)	4 Cost ABCD (×£1000)	5 If B is at (m depth)	6 Pipe size (mm)	7 Cost BC (×£1000)	8 Cost BCD (×£1000)	9 If C is at (m depth)	10 Pipe size (mm)	11 Cost CD (×£1000)
2·0	750	28·2	115·3	2·5	975	41·7	87·1	3·0	900	45·4
			112·2		900	38·4	84·0	3·5	825	45·6
			109·9		825	32·5	81·7	4·0	825	49·2
			111·6		750	32·0	83·4	4·5	825	51·4
			113·6		750	32·0	85·4	5·0	900	53·4
			120·8		750	32·0	92·6	5·5	900	60·6
			129·4		675	29·0	98·7	6·0	975	69·7
	675	21·7	115·4	3·0	975	48·1	93·7	3·5	825	45·6
			112·9		900	42·0	91·2	4·0	825	49·2
			109·1		825	36·0	87·4	4·5	825	51·4
			112·5		750	37·4	90·8	5·0	900	53·4
			119·7		750	37·4	98·0	5·5	900	60·6
			127·9		675	36·5	106·2	6·0	975	69·7
	675	21·7	119·0	3·5	975	48·1	97·3	4·0	825	49·2
			118·7		900	45·6	97·0	4·5	825	51·4
			116·1		825	41·0	94·4	5·0	900	53·4
			122·5		750	40·2	100·8	5·5	900	60·6
			131·6		750	40·2	109·9	6·0	975	70·7

600	20·8	129·7	4·0	975	57·5	108·9	4·5	825	51·4
		122·2		900	48·0	101·4	5·0	900	53·4
		126·3		825	44·9	105·5	5·5	900	60·6
		134·2		750	43·7	113·4	6·0	975	70·7
600	20·8	131·7	4·5	975	57·5	110·9	5·0	900	53·4
		134·8		900	53·4	114·0	5·5	900	60·6
		137·2		825	46·7	116·4	6·0	975	70·7
600	20·8	145·1	5·0	975	63·7	124·3	5·5	900	60·6
		147·5		900	56·0	126·7	6·0	975	70·7
525	26·1	159·5	5·5	975	63·7	133·4	6·0	975	70·7

the question 'If the depth at C is 3·0 m (or 3·5 m, 4·0 m, etc.) what is the best pipe size to use for length CD?'. The answer is given in column 10 and the corresponding cost in column 11. Similarly the situation from B onwards may be considered, for example by asking the question 'If the depth at B is 2·5 m what is the best combination of pipe sizes to use for BC and CD?'. The answer is that the lowest combined cost BCD is £81 700 with 825 mm pipes in both BC and CD, as shown in the third row of the table. The same calculation procedure is extended backwards to the length AB, and a cumulative cost figure for ABCD entered in column 4. Looking down column 4 it can be seen that the lowest total cost for the whole length ABCD is £109 100 with pipe diameters of 675 mm for AB, and 825 mm for BC and CD.

While in this case the whole table has been set out for completeness, it can be seen that many of the rows are superfluous. For example, for length AB a 675 mm pipe can carry the flow with a fall of less than 1·0 m but a 600 mm pipe needs a fall of more than 1·5 m. This means that where the available fall is only 1·5 m the larger pipe must be used, but cannot possibly offer lower costs than the previous block which showed a fall of only 1·0 m for AB. Hence the third block in the table could be omitted. Similarly the fourth, fifth and sixth blocks all use 600 mm pipe but of these the fourth must give the lowest costs; the fifth and sixth could therefore be omitted. In general, therefore only one depth increment for length AB need be considered for each pipe size AB, namely the lowest increment.

It is of interest to note that the solution of this problem gives sizes of 675 mm, 825 mm and 825 mm for the three pipe lengths. A preliminary inspection of the flow and cost figures might suggest that 600 mm, 750 mm and 900 mm would offer a sensible combination with increasing flows, but this would cost considerably more than the combination chosen above. Refinement of the solution is possible however, since the use of 0·5 m depth increments will mean that some depth is effectively wasted; this can be put right by taking the solution selected and working out the exact falls to achieve the required flows. Errors introduced by the use of 0·5 m depth increments could lead to the adoption of a solution other than the optimal one, but it is not likely to be very different from optimum.

Designers may feel that this method is not appropriate in many practical cases, since they are not in possession of cost data at the time of design. Even if exact costs are not known, it should be possible to give relative costs of the right order to enable a reasonable solution to be found.

References

1. R. W. Metzger, *Elementary Mathematical Programming*, Wiley, New York, 1958.
2. E. L. M. Beale, *Mathematical Programming in Practice*, Pitman, London, 1968.

3. E. G. Bennion, *Elementary Mathematics of Linear Programming and Game Theory*, Michigan State University, 1960.
4. K. E. Boulding and W. A. Spivey, *Linear Programming and the Theory of the Firm*, Macmillan, London, 1960.
5. F. S. Hillier and G. J. Lieberman, *Introduction to Operations Research*, Holden-Day, San Francisco, 1967, pp. 181–4.

6   Cost Models in Plant Location

In many decision-making situations we are concerned with optimisation in some form, and this very often comes down to cost minimisation. Many of the techniques in this book, and the particular examples quoted, are in fact concerned with cost minimisation. In most cases these are applications of decision theory, linear or dynamic programming, game theory or other particular techniques familiar to operational research workers. As has been discussed, all these techniques depend upon the creation of some form of model of the situation or problem under consideration. Often the use of a model alone is adequate for the solution of a problem, and in this chapter two examples are discussed in which the well-known operational research techniques are left aside in favour of simple graphical models. The first of these is called here 'Least-cost Route', and concerns the various decisions that have to be taken in setting up extraction plant for a particular mineral. While the case described relates to a specific example, there is no reason why the same technique should not be used in a wide range of problems. The method is based on simple mathematical techniques for finding the shortest path through a network, and is discussed in the subsequent paragraphs.

The method described here is somewhat similar to the lay-out of decision trees referred to in chapter 9. The essential difference is that in this case the alternative paths to be followed are the result of positive decisions, and do not have a chance element, as in the case of decision trees.

The second problem described in this chapter is concerned with the various decisions to be taken in the establishment of a ready-mixed concrete depot in an urban area. Two of the major decisions to be taken are the location and capacity of the plant, and an evaluation of the factors involved can be made using a form of contour map. The method described in the second part of this chapter is called here 'Plant-location Model'.

Least-cost Route

This case is based on a very simple application of the classical theory of graphs.[1,2] One particular aspect of the theory of graphs relates to a means for calculating the shortest path through a network and can be illustrated by reference to the simple diagram in figure 6.1.

Here the objective is to find the shortest path from a to b taking account of the length and direction of each link or arc in the network. While it is fairly

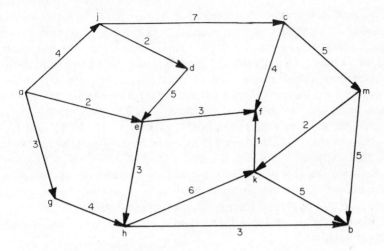

Figure 6.1   Finding the shortest path through a network

simple to spot the solution in this particular example, there is a rigorous method to find the answer, and this is as follows.

Step 1   Compare length *aj*, *ae*, *ag*, and select shortest, *ae* = 2, mark it in
Step 2   Starting from a or e what is next nearest point?
          Compare *aj*, *aef*, *aeh*, *ag*, select shortest, *ag* = 3, mark it in
Step 3   Compare *aj*, *aef*, *aeh*, *agh*, select shortest, *aj* = 4
Step 4   Compare *ajc*, *ajd*, *aej*, *aeh*, *agh*, select shortest, *aef*, *aeh* = 5, mark it in
Step 5   Compare *ajc*, *ajd*, *aehk*, *aehb*, select shortest, *ajd* = 6, mark it in
Step 6   Compare *ajc*, *ajde*, *aehk*, *aehb*, select shortest, *aehb* = 8
This then locates the required shortest path.

This method can clearly be used to find the shortest route on a road map where the arc lengths are quoted in miles, but it may equally well be used to find the quickest route when the arc lengths are given not in distance but in time. It is not difficult to draw an analogy with this in industrial production where the points or vertices of the network represent intermediate states of the material or product and the arcs or links represent the processes necessary to transform the product from one state to the next. In such an analogy the arc lengths may again be measured in time if it were desired to find the quickest way of coverting a raw material at point a, to a finished product at point b, or more usefully it may be that arc lengths are quoted in units of cost, so that the cheapest way of converting material a to product b can be found. This particular type of application is referred to as the *least-cost route*. The following case illustrates an application of this method but is so simple that it could be solved directly by comparison of alternatives.

On a large civil engineering contract a decision has to be taken about the method of concrete placing to be used. The main choice is between pumping directly and placing by crane. If the crane is used, the concrete must be brought within reach of the crane, either by pumping it to a fixed hopper or by tipper lorry. The alternatives are set out diagrammatically in figure 6.2. It is possible to calculate the cost per cubic metre of each of the proposed stages, and these are indicated on figure 6.3. The procedure is then to compare the costs of paths from box 1, namely £1·75, £0·53 and £0·42, select the lowest, £0·42, and mark in this path. Second, find the next nearest point and compare 1–2–4 having a unit cost of £0·42 + £1·06 = £1·48, with 1–4 having a unit cost of £1·75. The lowest-cost route is found to be 1–2–4 which indicates that the cheapest method is to use lorries to transport concrete to a hopper, and thence place it by skips lifted by crane.

A more complex example that would not readily be solved by simple arithmetic follows.

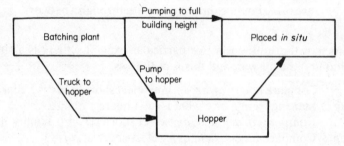

Figure 6.2   Alternative methods of concrete placing

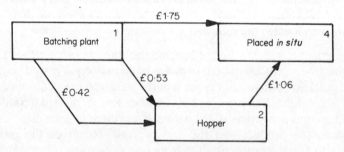

Figure 6.3   Costs of concrete placing

Mineral Extraction—Minimum-cost Layout

A recent study was made of a clay-like mineral (diatomite), which exists in considerable quantities beneath a loch on the Isle of Skye. There is a market for this mineral in a suitably processed form, and part of the study involved an economic assessment of extracting the mineral, processing it and delivering it to customers. In order to make this assessment it was necessary to determine plant capacities and types, and processing costs; an important factor in this was the decisions on digging methods and plant location. A number of alternative methods were available, and several decisions had to be considered.

The loch is five kilometres from the nearest road, located in a natural corrie in the hills, and is dammed up by a glacial moraine, see figures 6.4 and 6.5. The mineral is lying under about three metres of water and is itself about

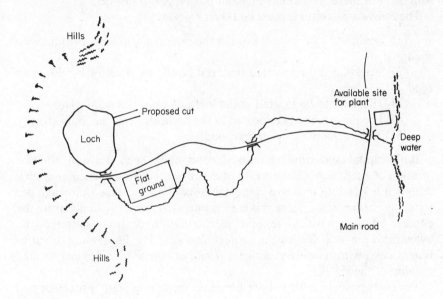

Figure 6.4   Site plan of mineral deposit

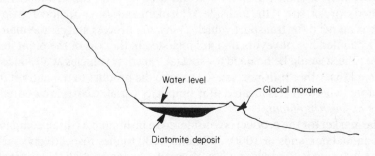

Figure 6.5   Vertical section through loch

nine metres in depth. It would be possible to drain the loch completely by making a cut through the moraine and then excavating the mineral in a semi-dry state; alternatively the mineral could be dredged in a wet state and stacked to drain. Two major plant items are required, namely a 'drier' and a 'processor' and these can be located either beside the loch or at the main road. In addition, intermediate and finished stocks are required to maintain a steady flow through the plant. There are capital costs associated with drainage of the loch, digging equipment, transport plant, drier, processor, road access to the loch, and storage buildings. Running costs are associated with most of these for labour, fuel and power, and transport.

The following decisions must be taken.

(1) Drain the loch in order to dig the material semi-dry, or otherwise dredge it wet.

(2) Stockpile the excavated material beside the loch or by the main road.

(3) The drier to be located at the lochside or at the main road.

(4) The processor to be located at the lochside or at the main road.

(5) Improvement of the access road.

Both capital and running costs depend upon each decision and the location of each item. Given a rate of depreciation and a particular annual output it is possible to assess the total costs of each production stage per tonne of final product. Since this case is intended primarily to illustrate the concept of 'least-cost route', the calculations have been simplified by assuming a ten-year life for the project, and ignoring discounting or variation of costs with time. In practice it would of course be necessary to take account of these.

The costs given in table 6.1 are based on the drying plant, processor and store being located at the main road; if they are on the site both capital and running costs would be ten per cent higher. Each vehicle could transport about 4000 tonnes per annum of finished product from the loch to the main road. The existing access road is adequate for site vehicles but would require improvement if road transport is to be used to take the product from a finished store on site at the lochside. If all plant and stores are located on site there is no need for transport vehicles between process stages, the material being handled by conveyors that are included in the cost of the plant items; further there would be no need for special transport vehicles at all, since it is assumed that other transport is used to take the product to the market from the store, wherever it is located. For simplicity variable costs are quoted *per tonne of finished output.*

The market for the product is considered to be limited to three companies, with annual demands of 1000, 3000 and 6000 tonnes respectively.

The first step is to prepare a flow diagram to represent all the alternative process stages through which the material may pass, and this is set out in

Table 6.1   Capital and running costs of mineral extraction

Item	Capital sum	Variable cost per tonne of finished output
Drainage of the loch	£95 000	£0·0
Dredging equipment	£50 000	£0·60
Drying plant	£55 000	£3·50
Processing plant	£45 000	£2·80
Digging equipment for either drained or dredged material	£22 000	£0·40
Vehicles to transport material from site to the road	£4 000 each	£0·15 each
Storage	£10 000	£0·10
Improvement of access road	£12 000	£0·05

figure 6.6. Each box is a vertex in the network and represents an *intermediate state in which the material may exist*; the first box indicates the material in its original condition, lying wet under three metres of water below the loch. There are two alternative paths from this point, either the loch may be drained, or the material may be dredged from beneath the water. If the loch is drained the next state of the material will be semi-dry still lying *in situ* on the bed of the loch. If, however, the material is dredged it would then be put in stock-piles beside the loch which would represent its next state. Box number 5 represents the material at its entry point to the dryer which is located at the site. Material may either be dug from its semi-dry condition in the bed of the loch, represented by the path from box 2 to box 5, or it may be dug from the stock-pile at the site, from box 3 to box 5. If, however, the dryer is located beside the main road, box 6, it is thought necessary to introduce an intermediate stock-pile at the main road, box 4. There are similarly boxes for the processor and store being located either at the lochside or by the main road and the arcs or arrows linking them represent the processes to be undertaken between the states of the material. The diagram can therefore be completed to show all the alternative routes through which the material may pass in being transferred from its original condition lying below the loch to its arrival at market.

The next stage is to calculate the costs of each stage of production, noting of course that costs will depend upon level of output. In general it would be necessary to examine the range of outputs associated with the upper and lower limits of the forecast demand. In this case it is known that there are only three potential customers and it is therefore possible to calculate costs for each level of output corresponding to the demands of any one, two or all three customers' needs. These costs are set out in table 6.2.

Table 6.2 Calculation of processing costs for mineral extraction

Market size		£1000	£3000	£4000	£6000	£7000	£9000	£10 000
Capital cost spread over 10 years and over each production level (£)								
Drainage	9500	9·5	3·16	2·38	1·58	1·36	1·05	0·95
Dredging	5000	5·0	1·67	1·25	0·83	0·71	0·56	0·50
Drying	5500	5·5	1·83	1·37	0·92	0·79	0·61	0·55
Processing	4500	4·5	1·5	1·12	0·75	0·64	0·5	0·45
Digging	2200	2·2	0·73	0·55	0·37	0·31	0·24	0·22
Vehicles	400/800/1200	0·4	0·13	0·10	0·13	0·11	0·13	0·12
Storage	1000	1·0	0·13	0·25	0·17	0·14	0·11	0·10
Road	1200	1·2	0·40	0·30	0·20	0·17	0·13	0·12
Add variable costs								
Drainage	0·00	9·5	3·16	2·38	1·58	1·36	1·05	0·95
Dredging	0·60	5·6	2·27	1·85	1·43	1·31	1·16	1·10
Drying	3·50	9·0	5·33	4·87	4·42	4·29	4·11	4·05
Processing	2·80	7·3	4·30	3·92	3·55	3·44	3·30	3·25
Digging	0·40	2·6	1·13	0·95	0·77	0·71	0·64	0·62
Vehicles	0·15/0·30/0·45	0·55	0·28	0·25	0·43	0·41	0·58	0·57
Storage	0·10	1·10	0·43	0·35	0·27	0·24	0·21	0·20
Road	0·05	1·25	0·45	0·35	0·25	0·22	0·18	0·17
On-site alternatives								
Drying		9·90	5·86	5·36	4·86	4·72	4·52	4·45
Processing		8·03	4·73	4·31	3·90	3·78	3·63	3·58
Store		1·21	0·47	0·38	0·30	0·26	0·23	0·22

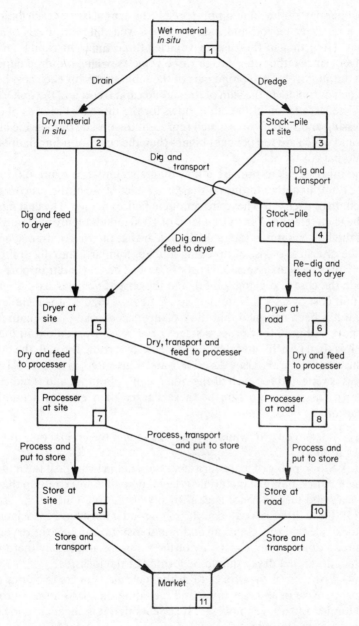

Figure 6.6　Alternative production flow patterns

Calculation of Costs

The upper part of the table simply spreads the capital cost of each major item over a ten-year period and over each of seven different levels of annual output. Note that in the case of vehicles three different capital sums are involved, since either one, two or three vehicles will be required depending upon output level. The second part of the table shows for each cost heading and each output level the sum of the spread capital cost and the unit variable cost. The last section of the table shows for the three cost centres of drying, processing and store figures which represent the on-site or lochside alternatives and which are ten per cent higher than the corresponding figures in the second part of the table.

The third step is to prepare a flow diagram similar to figure 6.6 for each level of output and enter into it the cost of each stage of the process that is, the length of each arc. These are set out in figures 6.7a–g. The cost figures in the diagram relating to an output level of 1000 tonnes (figure 6.7a) are drawn from the first column in the table. It is clear that the cost of drainage is £9·5 and the alternative cost of dredging is £5·6. Similarly the 'dig and feed to dryer' steps have an associated cost of £2·60. The 'dig and transport' stages include the cost of digging, £2·60, and the cost of vehicles, £0·55, giving a sum unit cost of £3·15. Note that no element of road cost is included here since it has been assumed that the existing access road is adequate for site transport. All the other process stages or arcs are worked out in the same way. It is then possible to calculate the least-cost route through the network by the method previously described and illustrated in figure 6.1a. This process is repeated for each of the other levels of output and is indicated in figures 6.7a–g. Examination of these figures then enables a number of conclusions to be drawn as follows.

(1) It is only worth while to drain the loch if the annual output is 9000 tonnes or more.

(2) Stock-piling of material is necessary at the lochside if the material is dredged. Stock-piling of material at the main road is an extra step that is not necessary and that does not appear in any of the solutions.

(3) If the dryer is located at the main road it must have associated with it a stock-pile that represents an additional cost. Location of the dryer at the main road does not appear in any of the solutions and it may therefore be concluded that the dryer should be located at the lochside.

(4) Location of the processor at the lochside appears to be marginally more expensive in all cases up to and including the 7000 tonne output. At 9000 tonnes and above, however, it appears that it is cheaper to locate the processor at the lochside.

(5) Improvement of the access road is only required if the finished product store is located at the lochside. The diagrams for 9000 tonnes and above indicate that this is marginally advantageous.

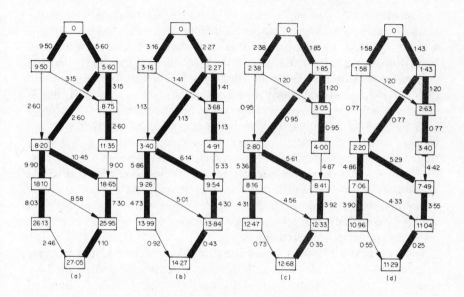

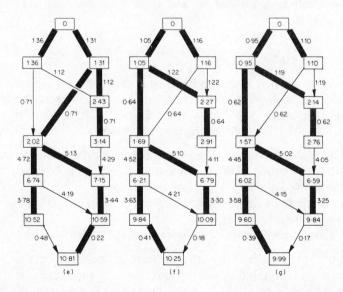

Figure 6.7   Cost network for various outputs (a) 1000 tonne (b) 3000 tonne
(c) 4000 tonne (d) 6000 tonne (e) 7000 tonne (f) 9000 tonne (g) 10 000 tonne
output

It is important to note that this method of analysis does not give any magic answer to the decisions that have to be made. What it does, however, is set out quantitatively the implications of each decision on the cost of the product. Many other factors would have to be taken into consideration in making the decisions and some of these are worthy of brief mention here. First, no assessment has been made of the probability of the market demand being at any particular level, and this would obviously be an important factor. Total capital availability would be another factor, and clearly drainage of the loch is a large capital sum that can only be justified if the output is high. Degree of risk must also be taken into account together with the residual value of equipment in the event of failure of the project. It is likely that dredging, digging equipment and vehicles would have a reasonably high residual value, the drying and processing plants being somewhat specialised would have a lower residual value but they in any case are not optional; expenditure on drainage of the loch or improvement of the access road would have virtually no residual value unless it could be shown that these offered some improvement to the area.

Some of these decisions are very sensitive to output level and to accuracy in estimates of cost. It would be feasible to examine this more fully by a sensitivity analysis of the various factors involved, but the volume of calculation would be considerable and would probably need to be carried out by computer.

Plant-location Model

The objective this time is to search for the best location for a supply depot, in this case a ready-mixed concrete plant in an urban area, and then to determine the most suitable capacity for the plant. There are many factors that will affect the decision, among the most important of them being

(1) the potential market for concrete in the area under consideration;

(2) the proportion of the concrete market that is likely to be supplied from ready-mixed plants;

(3) the share of the ready-mixed concrete market that an individual depot could expect to obtain;

(4) the expected capital and running costs of a plant.

Each of these factors entails a large study in itself, and here only brief mention is made of the first, second and fourth factors, since the method being illustrated is essentially concerned with the calculation of the market share of an individual plant and the way in which this affects both the location of the plant and its capacity. First, however, the other factors are discussed since it is necessary to put the problem of market share into proper context.

Potential Market for Concrete

First it is assumed that it is possible to define a geographical area within which the market must be examined. In the case of ready-mixed concrete it is safe to assume that there is a maximum radius beyond which it is not technically possible to deliver concrete, since the travelling time will exceed the maximum permissible mixing time of wet concrete. Having defined this area it is necessary to examine in detail the anticipated construction development that is likely to occur in the area. For most urban areas there now exists a long-term development plan that places ground into planning zones, and will determine the type of construction that might take place. The production of a forecast of new construction work within an urban area is not a simple task, and may not be achieved with great precision; however, some attempt at it is necessary.

The first analysis may simply be based on the floor area of buildings and some other convenient measure for roads, bridges, sewers and other major works. It should be possible to assess from past experience the quantity of concrete required in different types of construction, perhaps using a series of lines on a graph as shown in figure 6.8.

In the case of roads and other major works it is usually possible to obtain some idea as to whether these are to be elevated motorways, bridges, etc., and thereby to make an estimate of the volume of concrete required for their construction. From projected building programmes, it should then be possible to obtain a forecast of the annual demand for concrete within the geographical area that has already been defined.

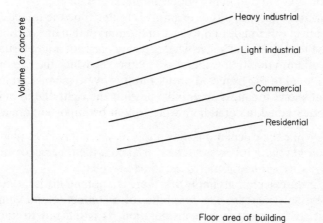

Figure 6.8   Concrete volume as a function of building type

Proportion of Concrete Market Supplied by Ready-mixed Plants

There are four main criteria that are taken into account in making decisions on whether concrete for a particular contract will be mixed on site or obtained from a ready-mixed depot. These are

(1) price
(2) quality
(3) space for plant and materials storage
(4) road accessibility with respect to traffic conditions

Most major contractors will have data available of the above form in relation to several contracts, and it should be possible to evolve a simple relationship between the total concrete volume on a contract and the proportion normally supplied by ready-mixed plants. This obviously will vary from one contract to another depending upon local conditions but there will be a general pattern in relation to the type of work; one such study revealed that about a quarter of all concrete was obtained from ready-mixed plants.

Plant Capital and Operating Costs

It is possible to estimate the capital and running costs of a concrete plant either from previous direct experience or by making a full assessment of the proposal to build a plant, taking advice from the plant suppliers and any other possible sources such as competitors, trade associations, suppliers and customers. The detailed approach to the preparation of such an estimate is not considered here but it should be remembered that many factors must be taken into account and many important decisions taken.

The primary purpose of this example is to illustrate the method by which an estimate may be made of the share of the market that one particular depot may hope to attract, and hence what capacity of depot might be considered. The calculations that follow indicate a method by which the size of the plant can be related to the expected profit, but it is by no means certain that the plant that shows the most acceptable profit is the right one to build. Other factors not studied in detail here which might be important are as follows.

(1) While the calculations may imply a reasonably large plant it is true that the larger plant involves a larger risk and if the forecast or calculations are in any way wrong this may lead to bad results.

(2) If the calculations imply that there is a potentially large market, this must inevitably increase the possibility of a competitor building another plant within the area under consideration. It is difficult to know how a competitor might react to a new plant being built in the area; if a large plant is built as a result of the study the competitor may then feel that the market in that area is well covered and that there is little point in trying to compete. On

the other hand he may feel that if one company has the confidence to build a large plant then the market potential must be considerable and he should also build a plant in the area. Under these conditions of doubt a cautious company may therefore build a plant that is not as large as might be indicated by the calculations and this fact may itself encourage a competitor to come in with a slightly larger, more economical plant.

(3) Physical space on the site under consideration and the acceptability of a ready-mixed plant to the local planning authority are of course also important factors.

(4) The availability of raw materials is of very great importance since this industry has a high material throughput. Location in relation to sources of supply of sand, gravel and cement and good site accessibility are therefore of paramount importance.

(5) The capacity of a ready-mixed concrete plant may be influenced by the company's trading responsibility to its customers. A company wishing to maintain good customer-relations may be prepared to build a plant with a capacity slightly in excess of the known maximum daily demand, in order to be able to satisfy all customers at all times. This will inevitably mean a relatively low plant utilisation and faced with this prospect a company might well take the view that it should build a plant with a capacity lower than the maximum daily demand and face up to the problem of having to persuade customers to accept a delay in delivery of concrete. It is difficult to justify excess plant capacity in these terms within the current structure of the construction industry. Many contractors' agents and foremen will state that it is very important for them to obtain good prompt service from ready-mixed plants, but it is unlikely that in practice they are prepared to pay a premium for such service. Site agents will rely upon establishing a good working relationship with the concrete-plant manager and thereby hope to obtain preferential treatment at times of high demand.

Market Share of the Individual Depot

The first step is to prepare a map of the area under consideration, mark on it the boundary determined previously and also mark on it the location and capacities of all competitors' plants. Such information is not difficult to obtain.

It is reasonable to assume that all plants within the area will act most of the time on a rational basis, that is, they will sell concrete at a price that is not less than the direct cost of materials and transport; this assumption is fundamental to the analysis that follows. The price will normally be in excess of this in order to make a contribution to overheads and to earn profit, but it is unlikely that any company will deliberately sell concrete at a price that is less than the sum of the direct costs involved in that particular sale. Since most ready-mix suppliers will be buying their materials from common sources, it should be possible to obtain the price paid for each type of material at each depot location. The cost of delivery of concrete can also be readily determined, especially since the common method of delivery of ready-mixed concrete is for truck drivers to own their vehicles and to be paid a sum for delivery that is directly proportional to the volume carried and the distance covered. If we now consider two depots A and B we can examine the minimum-price situation in the following way as illustrated in figure 6.9.

It is possible to find an intermediate point P, between depots A and B at which the minimum price that can be charged will be the same for the two depots, and this is given by the expression

$$M_A + kd_1 = M_B + kd_2$$

This clearly can be extended to consider points that are not on a direct line between A and B by adopting the method indicated in figure 6.10.

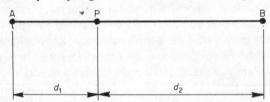

Figure 6.9   Transport costs of ready-mixed concrete—point of equal cost

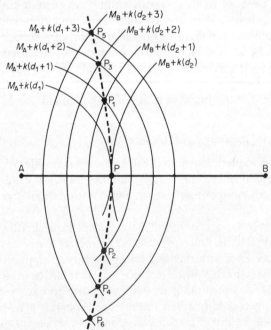

Figure 6.10   Transport costs of ready-mixed concrete—line of equal costs

A circle of radius d_1 is drawn with A as centre, and another circle of radius d_2 is drawn with B as centre. These two circles touch at the intermediate point between A and B at which the direct costs of the two depots equate, and have a value of $M_A + kd_1$ (or $M_B + kd_2$). A second pair of circles is drawn with centres A and B as before, but each with a radius one unit distance larger than before, that is, a radius $(d_1 + 1)$ centred on A and a circle of $(d_2 + 1)$ centred on B. These two circles intersect at points P_1 and P_2 which must also then be points of equal direct cost, that is, $M_A + k(d_2 + 1) = M_B + k(d_2 + 1)$. Further pairs of circles are drawn centred on A and B each time increasing the radius by one unit of distance, that is, $(d_1 + 2)$, $(d_1 + 3)$, etc. Where each pair of circles intersect two further points of equal cost are located. Joining up all the points $P_1, P_2, \ldots, P_n$ will give the line of equal cost between the two depots A and B. Every location to the left of this line will have a direct cost of supply that is lower for depot A than for B, and every point to the right of this line will have a lower cost for depot B than for A. It can be seen from the diagram that this is not a straight line but it has a curvature that is fairly slight unless there is a very large difference between the base cost of materials at A and at B, which in practice is most unlikely. If, however, there was a large difference between M_A and M_B such that $M_A - M_B > k(d_1 + d_2)$ this would mean that depot A would *nowhere* have a direct cost of supply lower than that of depot B, and it could not therefore compete profitably with depot B.

The same basic procedure can be adopted to cope with other factors entering the situation such as a third depot or a natural barrier such as a river. In the distribution of ready-mixed concrete, transport is paid for on a distance 'as the crow flies' except where it is necessary to make a considerable detour in order to cross a natural obstacle such as a river. In these cases the procedure for calculation of the delivery cost is to measure the distance from the depot to the bridge and then the radial distance from the bridge to the destination point.

In the following diagram, figure 6.11, three depots A, B and C are shown together with a river and the only road crossing-point. The base cost of raw materials at A, B and C is shown per unit output. The procedure previously described is used to draw the line between A and B that marks equal direct cost, and therefore divides the area to the south of the river into two parts; the one to the west being effectively a local market for A and the one to the east being effectively a local market for B. It is possible to draw a similar separation line between B and C taking account of the fact that the only river-crossing is at point X. In this case circles are drawn centred on C and X, the base costs at C being simply M_C but the base cost at X being $(M_B + kd_X)$ where d_X is the distance B to X. This takes account of the fact that for material to be supplied from depot B all vehicles must travel to X to cross the bridge before being able to spread out. The separation line between B and C divides the area into a local market around C to the west of the line; depot B

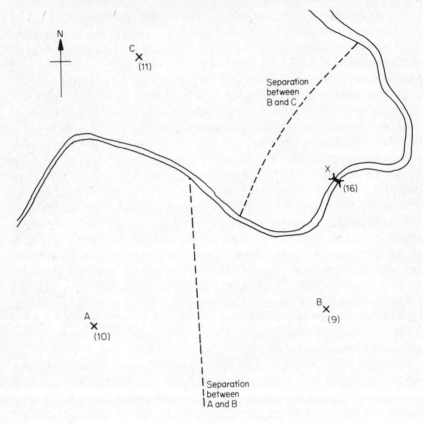

Figure 6.11   Separation of three home-areas with natural barrier

has the market south of the river already referred to, but in addition has a market to the north of the river and to the east of the line of separation from C.

The division of the whole area into zones described above does not necessarily mean that each depot will capture the whole market within its own local area. What it does mean is that the depot should be able to supply concrete to sites within its own area at a price equal to or lower than that of any of its competitors. This means that with careful pricing a depot should always be able to quote a price for concrete delivered within its own area low enough to beat all competitors. This does not ensure obtaining the order, however, because a competitor may act in an irrational way by quoting a price lower than his own direct cost, although this is unlikely and for the purposes of this analysis has been specifically excluded. There are of course other reasons why an order may not be placed with a particular depot; for example, a reputation for poor quality or failure to deliver on time, or other difficulties. These aspects are outside the present analysis but are of course extremely important and must not be neglected.

Pricing policy must depend upon many factors but one of them must be the anticipation of price that is likely to be quoted by a competitor. Using the foregoing analysis it is reasonable to conclude that at locations nearest to your own depot transport costs will be at a minimum, thereby giving a greater profit margin. At the same time at locations closest to your own depot it is likely that competitors will have a very high transport cost and therefore competitive prices will be relatively high; this means that prices quoted can be *higher* closer to the home depot, thereby permitting even larger profit margins. This aspect will be further explored later in the chapter.

Having examined the way in which a model can be drawn to represent the competitive position of a number of ready-mixed concrete depots within a given area, we may use this model to examine the potential for a new depot to be built within the area.

It has already been stated that the first step must be to produce a map of potential demand for concrete over the foreseeable future, together with an assessment of what proportion of this demand for concrete is likely to be supplied by ready-mixed concrete depots. Onto this map can then be plotted existing depots together with information on their base costs and separation into 'home areas' as outlined above. It would then be possible to produce an elegant theoretical solution which would suggest the ideal location for a new depot within the area at a point where density of demands and remoteness from existing depots showed the maximum potential market for a new depot. If this analysis were carried out in any existing urban area it is likely that such an ideal location would be a totally impractical one, since it would conflict with site availability, site cost and planning consent. The approach can however, be used to compare alternative sites that are available and this has been done in figure 6.12.

In the area under consideration there are three existing depots A, B and C and three possible sites are available for a new depot at P, Q and R. Assuming that the base cost of each depot is 10 units it is possible to consider each site in turn and to determine for each site a potential 'home area'. Following the procedure exactly as before a separation line can be drawn between depot sites A and P and between P and B, and with the river on the north this then bounds a 'home area' for depot P. An examination of the forecast market in that area will then give an estimate of the potential market for a depot built at location P. Similar lines can be drawn in the case of sites Q and R and hence potential demand for these two depots can be estimated. Note that depot Q does include within its home area a small area to the south of the river near the bridge, but depot R is restricted entirely to an area north of the river. It is also worth noting that although depot

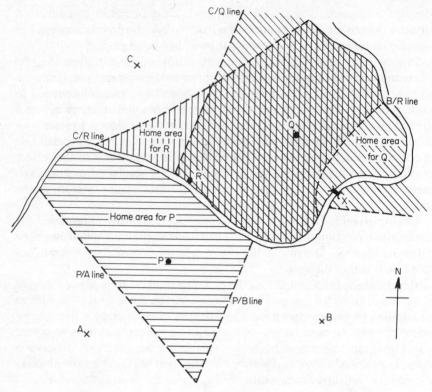

Figure 6.12   Home areas for three new depot locations

locations P and R are geographically quite close, their home areas do not overlap since the river intervenes and they are both at a considerable distance from the only river-crossing (X). This would mean that if a company were considering establishment of a ready-mixed concrete depot at P, Q or R the decision to build at P would be quite independent of the decision to build at Q or R, and it would be possible to open two depots, one at P and one at either Q or R without much likelihood of their interacting. Examination of the diagram, however, shows that depots Q and R would largely serve the same area and other factors might therefore become predominant in choosing between the two potential sites.

The method outlined in the preceding paragraphs enables a comparison to be made between different potential sites on the basis of the forecast market share that can be expected. It does not, however, give an estimate of the profit that may be expected, and in order to make a true investment appraisal in relation to the establishment of a new depot it would be necessary to make forecasts not only of the volume of output but also of its expected profit. A previous paragraph suggested that close to a depot costs are relatively low and prices relatively high, thereby permitting maximum profits. The next stage of the analysis is to examine this in a quantitative way and this can be done using price and profit contours as illustrated in the following paragraphs.

Assume that we are now considering the depot at location R as defined in the previous diagram. We have defined the area within which depot R should be able to compete profitably, and we now want to find what the profit might be resulting from sales within that area. The first step shown in figure 6.13 is to draw a series of concentric circles centred on C, R and X with radius intervals of one unit of distance. Only the part of the circle that falls within the home area of depot R need be drawn. Each circle can then be given a contour number, that number being equal to the base cost of material at the relevant depot plus the transport cost over the distance; for example, location 1 shown on the diagram lies outside contour C_{17} but inside contour C_{18} and hence the cost of supplying material from depot C to location 1 would be 18 units. Location 1 lies between contours B_{23} and B_{24} and hence the cost of supplying material from depot B to location 1 would be 24 units.

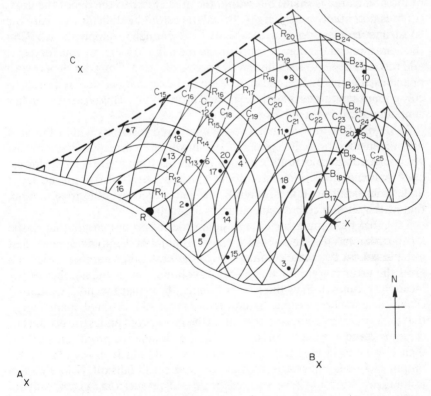

Figure 6.13 Cost contours for depots B, C, R and customer locations 1 to 20

From these two figures the manager of depot R can deduce that the minimum cost to a competitor of supplying concrete to location 1 is 18 units and therefore the minimum price that they are likely to quote is 18 units. At the same time he can see that location 1 lies within the contour R_{17} and he can therefore supply at a cost of 17 units. His gross margin for supplying to location 1 would therefore be $18 - 17$, that is, one unit. Location 2 lies within the contour C_{19} and within the contour B_{23} and hence the lowest competitive price that is likely to be quoted is 19 units. Location 2 however lies within the contour of R_{12} and hence depot R can supply at a cost of 12 units, giving a margin of $19 - 12 = 7$ units. Similarly location 3 lies within the contour C_{24} and within contour B_{19} and the lowest competitive price could therefore be quoted by depot B at 19 units. Depot R however, can supply at 17 units giving a gross margin of $19 - 17 = 2$ units. Evaluation of the profit potential can therefore be carried out by locating each forecast demand point on this contour map and thereby finding the gross profit that can be expected for depot R.

From this analysis it is clear that the gross profit obtainable in many parts of the home area is small, but within the area closer to the depot the gross profit is potentially much larger. It may therefore be difficult to justify the construction of a depot with a capacity large enough to supply the whole of the home area, and it may be more profitable to build a slightly smaller depot and restrict supply to a smaller, more profitable area. Consider the forecast demands at points 1, 2, 3, etc., in relation to the capacity of plant required, its consequent capital cost and the expected profitability. This is carried out in a simplified form in the tables 6.3 and 6.4.

Table 6.3 lists in column 1 twenty numbered locations within the area under consideration. In column 2 is shown the corresponding forecast volume in hundreds of units. Column 3 picks out from figure 6.13 the lowest possible competitive price and column 4 picks out the base cost for the depot under consideration, that is, depot R. Column 5 then gives the gross margin, that is, the difference between columns 3 and 4.

Table 6.4 rearranges the data in rank order of profit margin, that is, the location that has the highest profit margin is number 5 and this appears first in table 6.4. In this table column 1 gives the rank order number, column 2 gives the gross margin for that location, column 3 gives the number of the location, column 4 gives the forecast demand for that location. Column 5 shows the cumulative capacity required to meet all location demands up to that rank order number, column 6 gives the gross profit and is the product of columns 2 and 4. Finally column 7 gives the cumulative profit. This data is then presented in graphical form in figure 6.14, which shows that as the output increases the rate of increase of gross profit falls off. Note that this is not the relationship between gross profit and output from a given plant but is the relationship between gross profit and total market within a given area.

Table 6.3   Gross margins for customer locations

1 Location	2 Volume (units $\times 10^2$)	3 Lowest competitor	4 Depot R cost	5 Gross margin	6 Gross profit
1	1	18	17	1	1
2	2	19	12	7	14
3	4	19	17	2	8
4	6	20	15	5	30
5	4	21	13	8	32
6	3	18	14	4	12
7	2	15	14	1	2
8	2	21	19	2	4
9	1	21	21	0	0
10	3	23	22	1	3
11	3	21	18	3	9
12	4	18	16	2	8
13	5	17	13	4	9
14	2	21	14	7	14
15	2	21	15	6	12
16	4	17	12	5	20
17	1	19	14	5	5
18	4	19	17	2	8
19	7	17	14	3	21
20	2	19	15	4	8

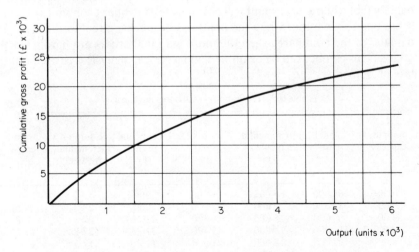

Figure 6.14   Gross profit with increase in plant output size

Table 6.4   Rank order profit margin for customer locations

1 Rank order	2 Margin	3 Location	4 Required capacity	5 Cumula- tive capacity	6 Gross profit	7 Cumula- tive profit
1	8	5	4	4	32	32
2	7	2	2	6	14	46
3	7	14	2	8	14	60
4	6	15	2	10	12	72
5	5	4	6	16	30	102
6	5	16	4	20	20	122
7	5	17	1	21	5	127
8	4	6	3	24	12	139
9	4	20	2	26	8	147
10	4	13	5	31	20	167
11	3	19	7	38	21	188
12	3	11	3	41	9	197
13	2	3	4	45	8	205
14	2	12	4	49	8	213
15	2	18	4	53	8	221
16	2	8	2	55	4	225
17	1	10	3	58	3	228
18	1	7	2	60	2	230
19	1	1	1	61	1	231
20	0	9	1	62	0	231

It can be seen that it is hardly worth while building a plant large enough to supply all demand within the 'home area' since the more remote locations offer very small margins if any. There is therefore a conflict to be resolved, namely that while a large plant will normally tend to offer economies of scale it has been shown that when the plant has a capacity equal to the forecast demand in the home area then the more marginal orders are hardly worth taking. This conflict can be examined and a typical analysis is carried out in table 6.5.

Table 6.5   Profitability related to plant capacity

1 Output capacity	2 Capital cost (£)	3 Incremental capital cost (£)	4 Gross profit (£)	5 Incremental profit (£)	6 Return on incremental investment (%)	7 Overall return (%)
1000	65 000	65 000	7 200	7200	11·1	11·1
2000	82 000	17 000	12 200	5000	29·4	14·9
3000	97 000	15 000	16 300	4100	27·3	16·8
4000	111 000	14 000	19 400	3100	22·2	17·5
5000	123 000	12 000	21 500	2100	17·5	17·5
6000	135 000	12 000	23 000	1500	13·8	17·1
7000	145 000	10 000	23 100	100	1·0	15·9

In this several different plants of capacity 1000–7000 units are shown together with their associated capital costs, which do not increase proportionately with output. Column 3 in this table shows the capital cost difference between each size of plant and one size smaller. Column 4 shows the gross profit that may be expected from each size of plant, drawing data from table 6.4 and figure 6.14. Column 5 shows the incremental profit for each size of plant, that is, the additional profit earned by that size of plant compared with a plant one size smaller. Column 6 shows the percentage return on each increment of investment, that is, column 5 as a percentage of column 3.

It is clear that the return on investment for a very small plant is likely to be low since it is probably operating in an inefficient way, which more than offsets the fact that most of the profitable section of the market can be met by a small plant. Building a 2000-capacity plant requires an additional £17 000 capital cost but at the same time produces an additional profit of £5000. It could therefore be said that the incremental return on investment of that additional £17 000 is at a rate of 29·4 per cent. Similarly the expenditure of a further £15 000 to increase the output capacity to 3000 units would bring in a further profit of £4000, that is, a rate of return of 27·3 per cent. Examination of column 6 however shows that the rate of return on subsequent increases of capital expenditure tends to fall off and it may be decided that once this rate drops below 17 per cent, or some other appropriate figure, it is not worth while carrying out further expansion of the plant. If a figure of 17 per cent is a reasonable rate for this type of project this would imply that the plant should be limited to a capacity of 5000 units, which would then only be able to supply 14 of the locations originally listed. Column 7 in table 6.5 shows the overall percentage return of each size of plant as obtained by taking column 4 as a percentage of column 2. Note that at a capacity of 5000 units the percentage return is 17·5 per cent and is therefore satisfactory. At a capacity of 6000 units the overall percentage return is still in excess of 17 per cent and would therefore appear to be acceptable, but it can be seen that the incremental return in taking the plant from 5000 to 6000 units shows a figure of only 13·8 per cent. This would indicate that it is preferable to restrict the plant to a capacity of 5000 units.

This analysis has shown a method of deciding the planned output capacity of the projected new depot at location R. There will be many other criteria to take into account; for example, site cost, accessibility, immediate surroundings, types of plant available, and so on, but these are not considered in this analysis. It is, however, possible to use the contour map (figure 6.13) in subsequent management of the plant at depot R once it has been established. The contour map can obviously be used to make an assessment of the minimum price that a competitor is likely to charge for delivery of concrete to a site at a specific location within the area. This can be read off directly from the contours without the necessity for carrying out calculations on each occasion that the depot manager is asked to quote for a supply. There is, of course, no reason why the contours should not be extended into the adjacent areas, which may be regarded as the home areas for B and C depots. This could be helpful in a situation where, say, C depot is known to be heavily committed and is therefore likely to be quoting relatively high rates in areas where its direct costs are high. With this information it is possible to find the areas in which depot C has a relatively small advantage over depot R and in which there are therefore reasonably good prospects of picking up marginal orders, if these are needed to keep the output from depot R up to capacity.

From figure 6.13 it is possible to produce another set of contours, namely those of profitability. These contours are shown in figure 6.15, which could serve as a useful basis for making positive efforts to obtain new business for the depot. It is obviously desirable to concentrate the selling effort of depot R in those areas in which profitability is seen to be highest, and it is worth noting that the highest profitability does not occur exactly in the immediate location of depot R. Where the potential profitability is relatively high, it is not only possible to beat competitors B and C on price, but it may be that such areas offer an opportunity for price reductions that would extend the category of business which goes to ready-mixed plants. Inspection of figure 6.15 will show that the separation line between home areas is simply the zero

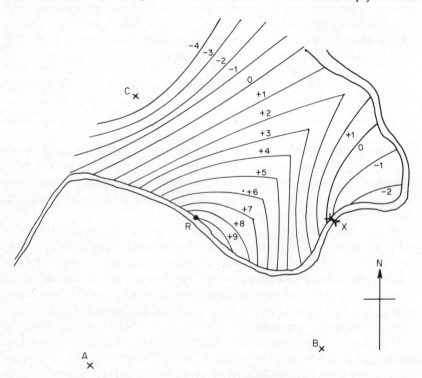

Figure 6.15   Profit contours for depot R

cost-difference contour on the system of contours. In the particular case considered the area is well bounded by a river, but the method could obviously be extended to cover a very much larger area. However, this is unlikely to be a realistic situation since even if there is not a physical boundary such as a river, it is possible that there is an effective boundary to the potential market area where planning zones change from either industrial or commercial or residential zones within the market area to agricultural or other green-belt areas which obviously offer virtually no market.

References

1. R. G. Busacker and T. L. Saaty, *Finite Graphs and Networks*, McGraw-Hill, New York, 1965.
2. L. R. Ford and D. R. Fulkerson, *Flows in Networks*, Princeton University Press, 1962.

7 Capital-investment Appraisal

Many of the techniques used in the appraisal of capital investment have been evolved by large capital-intensive industries. The chemical industry, for example, which typically has to invest large sums of money in highly complex plants that have a relatively limited life, has led the way in the development of methods of investment appraisal that are thought to be more soundly based and more realistic than traditional simple methods. The chemical industry is, of course, a major customer of the construction industry, others being property-development companies, manufacturing industry, local and central government. The attitude to return on capital in these various industries differs widely depending on the nature of the investment. A company building a chemical plant that has a limited life, will seek not only to recover the cost of the construction in a short period but will also be seeking to make an acceptable profit from the investment. In the area of property development it is unlikely that the developer will be seeking to recover the original capital cost since the value of property such as shops and offices will almost certainly continue to appreciate. In these circumstances the developer may be happy to accept a relatively low return on his investment in anticipation of capital appreciation, and in fact there have been many examples where virtually no return on capital has been achieved, for example, in the case of office blocks that have been allowed to remain empty in central London.

Investment by local and central government in such works as roads, bridges, water supply, effluent treatment, schools, colleges and hospitals cannot be assessed in the same way as commercial investment. It is not usual to consider the return on investment in a motorway or school, but it is more appropriate to consider value for money in social terms by such techniques as cost–benefit analysis. The method of appraisal to be used in capital investment is therefore very much dependent upon the view of the investor. While they may be of extreme interest in relation to the future prosperity of the construction industry, the methods of assessment employed by the property developer and government are of no immediate relevance to this chapter. The methods described below are more closely allied to those applicable within the chemical industry. This is because the form of investment is very similar; that is, in the construction industry capital investment on the part of contractors usually relates to the purchase of plant and equipment that has a limited life as a profit-earning potential. The methods described below are not exclusively relevant to the purchase of plant and may readily be extended to the consideration of investment in subsidiary companies, material sources such as quarries and other longer-term investments, but full discussion of these is better left to the professional accountants. Leaving aside therefore the more indeterminate cases of investment appraisal there are broadly two types of investment that we shall consider here, and in fact the method of appraisal is virtually the same in both cases.

The first group concerns consideration of the investment made by a client company who is having an industrial plant built that has a limited life and is expected to earn a profit. The second case relates to the assessment to be made by a contractor who is considering the purchase of plant to be used in the construction of a project. These two situations are broadly similar since they both involve investment in plant with a limited life that is expected to yield a return on the investment.

Accounting Practices

Before proceeding with the description of methods of assessment it is worth discussing briefly some of the accounting practices relevant to the construction industry. There are two distinct but complementary functions of accountancy in a company. The first of these is essentially that of book-keeping and is concerned with ensuring that payments of cash into and out of a company are carried out correctly, that accurate records of all transactions are kept and that the statutory records required by company law are maintained. Added to this factual record-keeping is the responsibility for ensuring the liquidity of the company. These aspects of accountancy are often referred to as external accounting, since they relate to the external transactions of the company and provide information to external bodies such as government, banks and share-holders.

The second major function of accounting is concerned with the provision of the data for an information system that enables the management of the company to exert control over its operation. This was not originally a function of accounting but since the most convenient unit of measurement of the company's activities is in terms of money, it has fallen to the accountant to provide management-control information. This aspect of his work is therefore often referred to as internal accounting since it provides information for use within the company for the control of its activities. It differs from external accounting in a number of ways and does require a rather different attitude of mind. The first difference is that for internal accounting, speed is more important than accuracy. It would be quite unthinkable in the maintenance of company accounts that sums of money in and out of the business be rounded up or down to a convenient number; accuracy is always required and the books must be made to balance even if this takes a long time. On the other hand, for the purposes of the control of production within a company it is important to know very quickly if something is going wrong with production costs, so that corrective action may be taken. For example, if a contractor is constructing a large area of reinforced-concrete wall he may have a target cost of £2·50 per m² for the shuttering. If for some reason costs rise to, say, £4·20 per m², it is important that he should know this as soon as possible and it is quite irrelevant that the actual cost may be £4·30 per m². The recording of accounting data may therefore depend upon whether it is to be used for internal or external purposes, and it may be quite wrong to use data from the same source for both types of accounts. For example, accurate figures must be used in the calculation of wages and bonus payments, but it is not important to have such accuracy when calculating the average cost per unit of production for a particular week. There may be cases when it is positively misleading to use the same data source for both financial bookkeeping and for cost-control purposes.

Having briefly discussed the difference between internal and external accounting, and having pointed out that the two may require rather different attitudes of mind, it is now proposed to proceed with the examination of investment appraisal assuming it to be an internal-accounting function. Although major investments must be considered along with all other expenditure by accountants performing their external function of bookkeeping, we are concerned here with the use of internal accounting as an aid to management in making decisions about investments. In this context we shall therefore now examine the alternative ways of assessing investments.

Return on Investment

This method (sometimes abbreviated to ROI) simply expresses the profit as a percentage of the sum invested. This apparently simple concept is, however, complicated by the following factors.

(1) It is necessary to state whether the percentage has been calculated before or after the deduction of Corporation Tax.

(2) Return on investment is usually calculated on the basis of profit after allowing for depreciation but this must be clearly stated. This is further complicated by the fact that the depreciation allowed for tax purposes is often different from the depreciation that a company may wish to apply, especially in the case of relatively high-risk projects.

(3) It is difficult to quote a single figure for profit when it is quite likely that profit may well vary considerably from year to year. Typically in industrial production the first year may show relatively low profit while the system is being run in; it should then show relatively good profits for a few years but it is likely to show a fall in profits as plant becomes obsolescent.

(4) It is difficult to define the capital sum being invested when allowance has to be made for investment grants made available by central government.

(5) This method takes no account of the timing of the receipt of profits. Receipt of £1000 in a year's time may be of equivalent value to the receipt of perhaps only £800 today, due to the combined effects of inflation and the earning power of the money during the intervening year.

(6) The return-on-investment criterion takes no account of whether the profits will be earned for one year, two years, ten years or more, and may well yield the same answer in each case.

(7) The single percentage figure quoted gives no measure of the capital sum invested. If the money is being invested to *achieve a particular purpose*, for example, in a large tower-crane, rather than simply as an investment to earn profit, then the sum involved is most important. It may well be preferable to invest £10 000 at a ROI of 10 per cent than £50 000 at 12 per cent ROI, if the same technical objective is achieved.

Pay-back Period

The basic concept in this case is to calculate the number of years after which the cumulative profits from the investment will exactly equal that investment. In this case depreciation is ignored, but deductions must be made from the profits for tax and the grants should be deducted from the original capital sum invested. Some of the objections to the return-on-investment concept can therefore be overcome by this method but it is still open to criticism that no account is taken of the profitable life of the investment nor of the timing of cash payments, nor of the size of the investment, as in notes 5, 6 and 7 above. Most of these objections can be overcome by the use of the method known as discounted cash-flow, usually abbreviated to DCF.

Discounted Cash-flow (DCF)

There is no doubt that the DCF method of investment appraisal involves considerably more arithmetic than either the return on investment or pay-back period methods. It does, however, overcome most of the faults of these two methods and the additional arithmetic is therefore usually justified. The DCF method has its critics and does indeed have some shortcomings, but the way that it is used in this book is suggested as being the most appropriate method for a construction company wishing to appraise in detail its investments in plant and equipment. Here we are not faced with the problem of whether we should invest £100 000 in excavating machinery or whether we should put the money into shares or government stock. The decision may be between buying 30 large-capacity heavy earth-moving vehicles or 40 slightly smaller-capacity vehicles of a different make and type.

The two distinctive features of this method are first, that it is concerned with the actual cash flows in and out of a project, thereby avoiding arguments about rates of depreciation. Second, the method takes account of the timing of payments by discounting them. Some or all of the following cash-flow items may be included in a particular investment appraisal.

(1) *Out*—the cash payment for the plant or equipment bought. This may be a single payment at the time of delivery of the equipment or it may involve a series of phased payments. If plant is being constructed on a site, as would be the case in a processing plant, it may be that progress payments are made at intervals during the course of construction. Each payment would be treated as an individual cash-flow item made at a particular time. If plant is not being purchased outright but is being bought under a hire-purchase agreement, then there will be no initial capital sum flow of cash but a series of separate cash payments over the term of the hire-purchase contract. These payments will cover not only the capital sum but also an element of interest on the capital employed. In the case where the purchase of equipment is financed by bank loan, then the initial capital payment is made by the bank and does not enter the cash-flow calculation at the outset. Payments of interest during the payment of the loan, however, will be entered as cash flows and then the total capital sum will enter as a cash-flow item at the time when the loan is finally paid off.

(2) *Out*—anticipated major items of repair, updating or modification of the plant expected during its lifetime. This would not normally include the cost of maintenance and minor repairs which would be regarded as a running cost.

(3) *Out*—tax will have to be paid on the profits earned by the investment. In order to calculate what tax will be payable it is necessary to estimate profits of the investment after taking account of depreciation as allowed for in the tax regulations. Payment of tax normally takes place in the year after the profit to which it relates has been earned.

(4) *Out*—working capital associated with the operation of the plant concerned in the investment. This covers the cost of materials, labour and other consumables of the work in progress, that is, the value of partly finished goods and stocks of raw materials and finished goods.

(5) *Out*—running-in expenditure which includes all the costs associated with starting up a new plant and running it up to normal operation.

(6) *In*—grants receivable from government or other authority. Capital grants associated with new plant investment are usually received at the time of the initial purchase and may therefore be dealt with simply as a reduction in the capital sum originally paid. If, however, there is a difference in timing between payment of the initial capital sum and the receipt of grant then they should be entered into the calculations separately.

(7) *In*—profits earned from the investment. This is essentially the gross profit associated with the investment before deduction of tax or depreciation. It is the balance of the revenue account, that is, the surplus of income

over direct expenditure less an allowance for overheads directly associated with the particular plant. Tax is not deducted since this is treated as a separate cash-flow item and depreciation is not deducted since we are concerned with actual cash flows associated with a project, not the notional writing down of the value of the asset.

(8) *In*—estimate of the cash received for the scrap value of the plant at the end of its working life.

Discounting and Net Present Value

This is an essential part of the DCF method and is essentially the concept that £100 payable next week is worth less than £100 today. More than this—the further a cash-flow payment is into the future, the less it is worth relatively in terms of today's values. Consider the simple example of a company that has to make a payment of £1000 one year from today. The company could invest a sum of X pounds at an interest of 10 per cent such that in one year's time the capital sum plus the interest earned would equal the £1000 necessary to make the payment. We could calculate the value of X as follows.

$$X_1 \times \left(\frac{110}{100}\right) = 1000$$

therefore

$$X_1 = \frac{1000 \times 100}{110}$$

$$= 909$$

Thus a cash-flow payment of £909 paid today into an investment which yields 10 per cent would produce £1000 one year hence. We therefore say that the *present value* of £1000 payable in a year's time discounted at 10 per cent is £909. By a similar calculation we can find the present value of £1000 payable two years hence as follows.

$$X_2 \times \left(\frac{110}{100}\right)^2 = 1000$$

therefore

$$X_2 = 1000 \times \left(\frac{100}{110}\right)^2$$

$$= £826$$

Thus the present value of £1000 payable two years hence when discounted at 10 per cent is £826. It is obviously possible by this method to calculate the present value of sums payable at any particular number of years into the future when discounted at any appropriate discount rate.

The generalised formula for this calculation is as follows.

$$\text{Net present value} = K\left(\frac{100}{100+r}\right)^n$$

where K is the sum payable

r is the discount rate (per cent)

n is the number of years hence that the sum is to be paid

While this can be readily worked out for each item in a cash-flow calculation the process is simplified by the use of tables that give values for the factor $[100/(100+r)]^n$, and such a table is reproduced at the end of this book.

In order to illustrate the method of discounted cash-flow a very simple example in which grants, profits and tax are excluded is taken first. A man has sold some personal property and he is offered the choice of two alternative forms of payment. The first alternative A consists of the payment of £100 each year for the next six years, the first payment being due one year from now. The second alternative B is the payment of £500 immediately. Alternative A involves a greater total sum payable, namely £600, compared with only £500 under alternative B but the latter has the advantage that the whole sum is payable immediately. How might the man assess these two alternatives? If we assume a discount rate of 10 per cent the calculation could be carried out as follows.

Table 7.1   Net present value of £100 payable annually for six years

Year	Sum payable	Net present value
1	£100	$0{\cdot}909 \times £100 = £90{\cdot}9$
2	£100	$0{\cdot}826 \times £100 = £82{\cdot}6$
3	£100	$0{\cdot}751 \times £100 = £75{\cdot}1$
4	£100	$0{\cdot}683 \times £100 = £68{\cdot}3$
5	£100	$0{\cdot}621 \times £100 = £62{\cdot}1$
6	£100	$0{\cdot}564 \times £100 = £56{\cdot}4$

Sum present value £435·4

This calculation shows that the net present value of the six separate payments of £100 over the next six years is only £435 and therefore the immediate payment of £500 is preferable. A second simple problem introduces a further concept as follows.

In this example grants, tax, profits, etc., are again ignored and we are simply considering cash flows. An organisation is given the opportunity of investing £5000 now and will receive a payment of £1000 per annum starting twelve months from now for the next ten years. The problem is to decide whether the organisation should take the opportunity. As in the previous example the present value of the payments may be calculated as in table 7.2. In this example, however, the calculation is carried out twice, once for a discount rate of 10 per cent and once for a discount rate of 20 per cent.

Table 7.2   Net present value of £1000 payable for ten years

Year	Sum payable	Present value at 10% discount	Present value at 20% discount
1	£1000	£909	£833
2	£1000	£826	£694
3	£1000	£751	£579
4	£1000	£683	£482
5	£1000	£621	£402
6	£1000	£564	£335
7	£1000	£513	£279
8	£1000	£467	£233
9	£1000	£424	£194
10	£1000	£386	£162
	Sum present value £6144		£4193

Table 7.2 shows that if the discount rate is taken as 10 per cent the sum present value of the ten payments of £1000 is £6144. If we take the investment of £5000 as a cash outflow now we then say that the proposition has a net present value of £6144 − £5000 = £1144. However if we choose to discount at 20 per cent the sum of the present values of the ten payments of £1000 is only £4193. Again taking the £5000 investment as a cash outflow now, this shows that the net present value of the proposition is £4193 − £5000 = −£807. A decision on whether or not to accept the proposition then would depend upon the value of discount rate chosen. In this problem there must obviously be one discount rate that would yield a present value of the £1000 payments of exactly £5000. This is known as the *solution rate of return* and it is the percentage discount rate that gives a value of zero to the net present value of all cash flows in the investment. In other words it may be regarded as the rate which, if applied as a discount rate in the DCF calculation, will ensure exact repayment of all the cash outflows. The actual calculation of this solution rate of return involves fairly tedious arithmetic and it is usually simpler to use a few trial rates, plot these on a graph and then by interpolation find the solution rate of return. In the case of the problem just considered this works out at a value of 14·9.

Example of Plant-investment Appraisal

A company is assessing the acquisition of heavy earth-moving equipment for a specific large contract and wishes to consider two alternative propositions A and B that are both technically acceptable for the type of work to be undertaken. The facts to be considered are as follows.

Table 7.3   Plant investment example—base data

	Equipment A	Equipment B
Number of machines needed	4	3
Length of project (months)	20	24
Capital cost of each machine	£37 500	£63 333
Scrap value of each machine at end of project	£2 000	£6 667
Monthly profit earned on project	£10 000	£10 000
Major overhaul after one year, each machine	£8 000	£8 000

A number of assumptions are made in the analysis of this problem. The project starts at the beginning of a tax year and tax is payable one year after the completion of the year to which it relates. Tax is payable at 52 per cent. A government investment grant of 20 per cent is payable on the original capital sum, and the balance depreciated within the first year. Tax in this case is therefore due on the profits of the second year of operation only, and is paid at the end of the third year. The major overhaul cost is treated as a capital sum not eligible for grant, but chargeable against profits. Profits are assumed to accrue from monthly progress payments one month in arrears of completion of each month's work. When carrying out DCF calculations a discount rate of 2 per cent per month is used, which is equivalent to an annual rate of 26·8 per cent.

A calculation can be made on the basis of the traditional return on investment criterion as shown in table 7.4. There is little to choose between equipments A and B whichever form of ROI is used, but it is of interest to note that so many and varied values can be given to this percentage. Also to calculate an annual return on a project of such short duration is not very meaningful.

It is possible to calculate the pay-back period in a similar way, with the result shown in table 7.5. The pay-back criterion suggests that equipment A is marginally preferable, but this takes no account of the relatively high overhaul cost of equipment A, nor of its relatively low scrap value at the end of the project. The pay-back criterion also ignores what happens after the capital sum has been recovered, in this case equipment A will continue to earn profit for a further 4·8 months while B will earn for a further 6·4 months.

Table 7.4 Plant investment example—ROI

(All figures in £ × 10³)	Equipment A		Equipment B	
Capital cost	150		190	
Overhaul cost	32	182	24	214
Scrap value	8		20	
Grant received	30	38	38	58
Total expenditure		144		156
Total profits	200		240	
Less depreciation	144	56	156	84
Tax payable at 52%		29·12		43·68
net profits		26·88		40·32
ROI expressed as percentage of initial net cost (equivalent annual rate)	$\frac{12}{20} \times \frac{26\cdot88 \times 100}{120}$ $= 13\cdot44\%$		$\frac{12}{24} \times \frac{40\cdot32 \times 100}{152}$ $= 13\cdot25\%$	
ROI expressed as percentage of total capital outlay (equivalent annual rate)	$\frac{12}{20} \times \frac{26\cdot88 \times 100}{182}$ $= 8\cdot86\%$		$\frac{12}{24} \times \frac{40\cdot32 \times 100}{214}$ $= 9\cdot42\%$	
ROI expressed as percentage of the average written-down book value (equivalent annual rate)	$\frac{12}{20} \times \frac{26\cdot88 \times 100}{79}$ $= 20\cdot16\%$		$\frac{12}{24} \times \frac{40\cdot32 \times 100}{105}$ $= 19\cdot14\%$	

Table 7.5 Plant investment example—pay-back period

	Equipment A	Equipment B
Net capital outlay	120 + 32	152 + 24
Gross monthly profit	10	10
Pay-back period in months	152/10 = 15·2	176/10 = 17·6

The DCF calculation is set out in table 7.6, showing the flows of cash in and out of the two alternative projects. The first row shows the full capital cost as an outflow and the grant as an inflow. The overhaul cost and scrap value recoveries are entered at the appropriate times. In the case of either equipment A or B the net present value of all future cash flows has a value not much greater than zero, and there may therefore seem to be little to choose between the two alternatives. However, it is interesting to plot a graph showing the net present value of all cash flows up to any particular point in time, that is, the cumulative present value of cash flows shown in the columns of table 7.6.

Table 7.6   Plant investment example—DCF

End of month	Discount factor at 2% p.m.	Equipment A				Equipment B			
		Cash out (£×10³)	Cash in (£×10³)	Present value (£×10³)	Cum. present value (£×10³)	Cash out (£×10³)	Cash in (£×10³)	Present value (£×10³)	Cum. present value (£×10³)
0		150	30*	−120·00	−120·00	190	38*	−152·00	−152·00
1	0·961		10	9·61	−110·39		10	9·61	−142·39
2	0·942		10	9·42	−100·97		10	9·42	−132·97
3	0·924		10	9·24	−91·73		10	9·24	−123·73
4	0·906		10	9·06	−82·67		10	9·06	−114·67
5	0·888		10	8·88	−73·79		10	8·88	−105·79
6	0·871		10	8·71	−65·08		10	8·71	−97·08
7	0·853		10	8·53	−56·55		10	8·53	−88·55
8	0·837		10	8·37	−48·18		10	8·37	−80·18
9	0·820		10	8·20	−39·98		10	8·20	−71·98
10	0·804		10	8·04	−31·94		10	8·04	−63·94
11	0·788	32	10	−17·34	−49·28	24	10	−11·03	−74·97
12	0·773		10	7·73	−41·55		10	7·73	−67·24
13	0·758		10	7·58	−33·97		10	7·58	−59·66
14	0·743		10	7·43	−26·54		10	7·43	−52·23
15	0·728		10	7·28	−19·26		10	7·28	−44·95
16	0·714		10	7·14	−12·12		10	7·14	−37·81
17	0·700		10	7·00	−5·12		10	7·00	−30·81
18	0·686		10	6·83	+1·76		10	6·83	−23·98
19	0·673		18†	12·11	+13·87		10	6·73	−17·25
20	0·660		10	6·60	+20·47		10	6·60	−10·65
21	0·647						10	6·47	−4·18
22	0·634						10	6·34	+2·16
23	0·622						30†	18·66	+20·82
24	0·610						10	6·10	+26·92
25									
36	0·490	29·12		−14·27	+6·20	43·68		−21·40	+5·52

* 20 per cent industrial grant

† Includes scrap value realised at end of project

Figure 7.1 shows that the net present value of cash flows to date has a lower value for *B* (dashed line) than for *A* (full line) for most of the duration of the project. This is a measure of the liability at any stage of the project and reflects the position if anything should go wrong and prevent its continuation. On this basis it might well be concluded that since *A* and *B* have effectively the same net present value for the project as a whole, than it would be prudent to select *A*, since if the project were to be stopped part way through for any reason, then alternative *A* shows a smaller net outflow of cash.

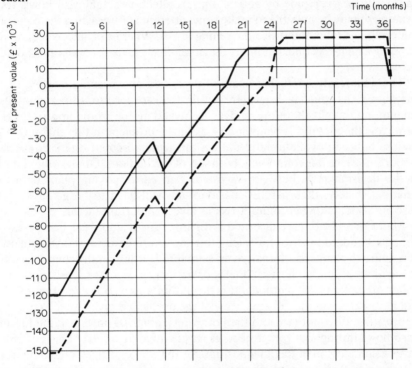

Figure 7.1   Net present value of cash flows in projects *A* and *B*

Since the net present value of all cash flows for the project is close to zero in both cases *A* and *B*, it may be concluded that the solution rate of return is very close to the 2 per cent per month (26·8 per cent per annum) used in the calculations.

The various criteria for assessment in this example have now given different answers to the problem of which equipment to select. All forms of the return-on-investment percentage showed no clear preference, while the pay-back period concept gave a preference for *A*. The DCF calculation showed little to choose between the two provided that the project ran its full duration, but indicated a preference for *A* since throughout the project the commitment was less.

Selection of Discount Rate

The selection of an appropriate discount rate is a subject open to great debate. Clearly the rate must be high enough at least to cover the yield that could be obtained by bank deposit or other secure investment. It may be argued that it should also exceed a notional mean equity yield since this offers an alternative for investment. The problem is confused in an inflationary situation, since if the value of the £ falls then a delay in its receipt represents a cost that must be taken into account. The difficulty of inflation can partly be overcome by anticipating its effect on actual cash flows, for example it may be possible to make some attempt at forecasting the costs of materials and labour and also the selling prices. Where a project is regarded as being a high-risk one, it is likely that a relatively high discount rate should be used.

The apparently confusing position described above could suggest that discount rates could be anywhere within the range 10 to 40 per cent per annum, and might therefore appear to invalidate the whole approach. If, however, the method is being used—as in the problem in this chapter—to choose between two alternatives, then the actual discount rate used is less important, since it is common to both sets of calculations. On the other hand if the decision rests between investment in earth-moving plant or in a building society, then careful consideration will have to be given to the discount rate, taking account of risk as a very important factor.

Many organisations concerned with investment appraisal have established a standard practice for calculating the solution rate of return, and use this figure to compare alternative investments. Because of the complexity of the arithmetic, many short-cut methods have been developed, using graphical or other empirical methods. It is the author's view, however, that this is not necessary, and that a better view of the problem is given by selecting an appropriate discount rate and then drawing a graph of present value for all cash flows throughout the project, as was done in figure 7.1. This not only gives a comparison of the final position for the whole project, but compares alternatives at all stages of the project indicating the degree of vulnerability to loss in the event of a premature end to the project. This is illustrated by the following simple example, in which two investments X and Y are compared. The cash flows of the two projects are set out in table 7.7, showing that in cash terms both projects have a net inflow of £700. However, since the timings of the cash flows are different the net present value for the two projects differs considerably.

Figure 7.2 compares the two investments with placing the whole capital sum in a fixed interest investment Z at 15 per cent per annum, repayable at the end of ten years. Z and Y are comparable when the whole term is considered, but Z is comparable with the inferior alternative X before the capital sum is repaid. Under these conditions there is no doubt that investment Y offers the best alternative.

Table 7.7 Cash-flow comparison of projects X and Y

End of year	Discount factor at 15% p.a.	Project X				Project Y			
		Cash out	Cash in	Present value	Cum. present value	Cash out	Cash in	Present value	Cum. present value
0		1000		−1000	−1000	400		−400	−400
1	0·870	300		−261	−1261	400	200	−174	−574
2	0·756		200	151	−1110	400	300	−76	−650
3	0·658		200	132	−978	100	400	+197	−453
4	0·572		200	114	−864		400	+229	−224
5	0·497		200	99	−765		300	+149	−75
6	0·432		200	86	−678		100	+43	−32
7	0·376		200	75	−603		100	+38	+6
8	0·327		200	65	−538		100	+33	+39
9	0·284		200	57	−481		50	+14	+53
10	0·247		400	99	−382		50	+12	+65
Totals		1300	2000			1300	2000		
net gain		£700				£700			

Table 7.8 shows in detail the present value over a ten-year period of investment Z with a capital sum of £1300 and a fixed interest rate of 15 per cent.

Table 7.8   Cash flow of investment in security Z

End of year	Discount factor at 15% p.a.	Cash out	Cash in	Present value	Cumulative present value
0		1300		−1300	−1300
1	0·870		195	+170	−1130
2	0·756		195	+147	−983
3	0·658		195	+128	−855
4	0·572		195	+112	−743
5	0·497		195	+97	−646
6	0·432		195	+84	−562
7	0·376		195	+73	−489
8	0·327		195	+64	−425
9	0·284		195	+55	−370
10	0·247		1495*	+369	0
	Totals net gain £650	1300	1950		

* Includes repayment of the capital sum

Sensitivity to Accuracy of Data

Capital-investment appraisal is a subject in which sensitivity analysis is particularly appropriate. Critics of DCF and its derivatives will often question the validity when data is not reliable, by asking such questions as 'What happens if the capital cost turns out to be 10 per cent higher than estimated?'. The method of sensitivity analysis sets out to answer this question *exactly*, rather than simply accept it as a criticism of DCF. It is appropriate to work out the net present value of a project for a whole range of possible variations from the conditions assumed in the original estimate. Among the factors that may be reviewed in this way are

Increase in capital cost of the project
Variation in running cost of the plant
Variation in revenue due to change in market rates
Effect on revenue of variation in plant utilisation
Delay in the completion of construction and hence delay in revenue
Plant life being greater or less than that estimated

The relative importance of these factors can be assessed for a particular project and presented in the form of a variation graph of the form shown in figure 7.3.

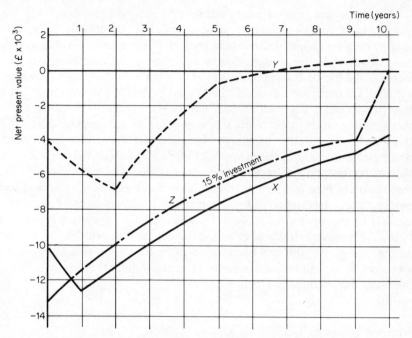

Figure 7.2 Net present value of cash flows in projects X and Y compared with investment Z

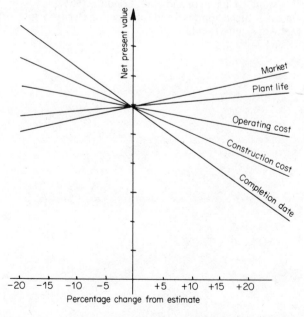

Figure 7.3 Variation of net present value with changes in input data

This figure not only provides answers to sceptical managers who are questioning the value of investment appraisal, but also provides a means of making decisions about details of the project. For example, a question might arise about a design change that would reduce the operating cost and increase the life of the plant, presumably at an increased capital cost and with a delay in the date at which the project begins to earn revenue. These four factors obviously interact and it would be relatively simple to assess their combined effect by reference to a graph of the form of figure 7.3. Before a final decision to adopt a modification was taken, however, it would be advisable to *evaluate* the net present value of the amended proposal, and then carry out further sensitivity testing of the amended project. The interaction of time and cost of construction requires special consideration, perhaps along the lines considered in chapter 3, and will of course vary very much from one project to the next. Delay in project completion is shown in figure 7.3 to have a dramatic effect on the net present value of the project, and in many cases this has been found to be the case.[3] This gives further emphasis to the importance of network analysis and other project-control methods, as described in chapter 2.

References

1. C. T. Horngren, *Accounting for Management Control*, Prentice Hall, Englewood Cliffs, N.J., 1970.
2. A. M. Alfred and J. B. Evans, *Discounted Cash Flow—Principles and some Short-cut Techniques*, Chapman and Hall, London, 1965.
3. P. A. Thompson, Project Appraisal and Engineering Decision, International Cost Engineering Symposium, London, 1974.

Part Three   Uncertainty and Risk

It is characteristic of many decisions in management that they have to be made before all the relevant facts are known; in other words decision-making under conditions of uncertainty. Chapter 8 deals with this topic, introduces the concepts of decision theory and utility theory, and quotes examples of their application in the construction industry. Single decisions are discussed in chapter 8; in chapter 9 the more complex situation of multi-stage decisions, that is, situations where a series of interactive decisions has to be made, is discussed. Finally chapter 10 considers the specific area of competitive tendering for contracts, where a continuing series of interacting decisions has to be taken, each decision probably being affected by its predecessor.

8 Decision-making under Conditions of Uncertainty

As was explained in chapter 4 it is convenient in an analysis of decision-making to differentiate between conditions of certainty and conditions of uncertainty. In this context certainty does not mean an *exact* knowledge of every detail relevant to the problem under consideration, but it does mean that we have a *reasonably good* idea of the value of all relevant factors. By contrast when we speak of uncertainty we mean that there are factors that may or may not arise or whose value may vary over a wide range; the essential feature of conditions of uncertainty is that there is an element of chance associated with these factors. The difference between certainty and uncertainty was illustrated by the water-tower design example described in chapter 4. Two other problems are now examined and serve to illustrate various aspects of decision theory.[1]

In the construction of a bridge over a river that is subject to flooding, a coffer-dam structure of steel sheet piling has to be built as part of the temporary works. This structure will only be in existence for a relatively short period of time during the construction of the bridge foundations and a decision has to be taken as to the height of the coffer-dam that will be required to keep out flood waters. It is obvious that the higher the wall the more expensive it will be, but also it will be safer. A lower wall will be cheaper but will bear some risk of being inadequate; how can the best height of wall be decided?

A second problem is one that faces a contractor who is bidding for a series of road contracts in a given area where there is a shortage of suitable fill materials. The contractor may have the opportunity to buy a local quarry that can supply suitable material, but he has to decide whether or not to buy the quarry before he has tendered for the contracts. How should he decide whether or not to take up the option on the quarry? This problem is discussed fully in chapter 9.

Before it is possible to examine these particular problems it is necessary to look briefly at some of the concepts that will be used.

Probability

In chapter 4 an explanation was given of the meaning of probability, the way in which it is used in this book in particular and in operational research studies in general. It is sufficient here simply to review the idea that when there are two or more possible outcomes of a particular situation then it is possible to assign a numerical value to the probability that each of these particular outcomes will arise.

Outcome or Pay-off

When a situation arises in which a decision has to be made and a particular outcome may result, it is usual to give a value to that outcome. This is sometimes referred to as pay-off; for example, if a contractor was successful in obtaining a contract at a tender price of £100 000 and incurred costs of £90 000 in the execution of that contract, the pay-off from the contract would be gross profit, that is, £10 000. If, however, his costs came to £95 000 his gross profit or pay-off would only be £5000, but similarly if he were able to complete the work at a cost of £85 000 the pay-off would be £15 000. A loss would be shown as a negative pay-off. Note that no account is taken here of company overheads, and normally these have to be recovered out of gross profit or pay-off.

Expected Value

This is a notional concept that gives a measure of the expectation of pay-off that may arise from a particular situation. This is obtained simply by multiplying the pay-off of each possible outcome by the probability that that outcome will arise, and then summing all these products. Assigning probabilities to the three different costs of the simple contract referred to above produces an expected value for the outcome of the contract as follows in table 8.1.

Table 8.1   Expected value of contract profit

Tender sum (T) (£)	Contractors cost (c) (£)	Profit $(T-c)$ (£)	Probability of this cost arising	$(T-c)p$
100 000	85 000	15 000	0·1	1500
100 000	90 000	10 000	0·6	6000
100 000	95 000	5 000	0·3	1500
			Expected value of profit =	9000

In this example the expected value of profit is £9000 but this does not mean that the profit arising from this particular contract will be exactly £9000 but simply that if enough contracts of this type are undertaken with this sort of probability distribution of costs, then on average the profit for such contracts would be £9000. Indeed in some cases it may be that the expected value of an outcome has a value that may never arise in an individual case. Consider for example, the very simple case of spinning a coin, where two people A and B agree that if the coin comes up heads, A will pay to B £1, but if it comes up tails, B will pay to A £1. Now the probability of heads is 0·5 and the probability of tails is 0·5, and if we calculate the expected value of a gain from A's point of view, it will be

$$0·5 \times (-£1) + 0·5(+£1) = 0$$

Note that in this expression a payment of £1 from A to B is written as $-£1$ from A's point of view and a payment of £1 from B to A is shown as $+£1$ from A's point of view. The expected value comes out at 0, which is the average outcome that may be expected if this game were played for long enough. It does not, however, represent the outcome of any individual game, the pay-off for which must either be a payment of £1 from B to A or a payment of £1 from A to B, that is, the pay-off must in any individual game be £1.

Temporary works—Coffer-dam

Using these general concepts it is now possible to consider a specific example. This concerns the construction of a coffer-dam of steel sheet piling with all its associated support structure, forming part of the temporary works for the construction of bridge piers in a river that is subject to both tidal and seasonal flooding. Part of the design of the coffer-dam consists of making a decision on its height in relation to the expectation of highest water level. In the construction of permanent works it might well be argued that the highest water level that could ever be anticipated should be the basis of design, but in the case of temporary works that may only last for a few weeks or months, it may be argued that some lower level would be adequate. Such a decision would imply that a risk may be taken in the construction of temporary works that is not acceptable in permanent works. If such a risk is involved it is of value to know how high that risk is and what would be the result if a flood in excess of the height chosen should arise. There are several factors to be taken into account; the height of the wall obviously affects its cost, not only because of the extra sheet piling involved but also the cost of the supporting structure and the obstruction effect on subsequent construction work. If there should be a high water level in excess of the coffer-dam height there would presumably be damage involved both to the temporary structure itself and to any permanent works completed. Should flooding of the works occur there would almost certainly be delays in the project and this may in turn have a cost element. There should be records available of tides and floods that would enable an estimate to be made of the probabilities of various flood levels arising. The information relevant to this example is given in table 8.2.

Table 8.2   Costs and probabilities in coffer-dam problem

Wall height above datum level (m)	3	4	5	6
Cost of construction of coffer-dam ($£\times10^3$)	18	24	40	44
Probability that one tide will exceed this height but not the next	0·5	0·3	0·05	0·05
Cost of damage repair, etc. ($£\times10^3$)	50	52	54	56

A more careful look into the probability distribution should really be taken, since it is possible that more than one high flood could occur during the construction period. In order to simplify this case, however, it is assumed that if one high tide does occur and cause damage to the coffer-dam then the result would be such that the contractor must adopt a different approach to the construction method in order to make up for time lost and would therefore not be susceptible to a second or subsequent high tide within the period. The cost of any alternative method of construction made necessary in this way is included in the figures for the costs of remedial action given in the table.

The first stage in the calculation is to work out for each height of wall the total cost that would arise if the highest tide falls within each of the ranges shown. It is thus possible to produce a cost matrix as laid out in table 8.3.

Table 8.3   Pay-off matrix for coffer-dam

Maximum flood within the range (m)		up to 3	3–4	4–5	5–6	over 6	Expected value of total cost
Probability that maximum flood is within this range		0·1	0·5	0·3	0·05	0·05	
Wall height (m) with corresponding costs, construction plus repair ($£\times10^3$)	3	18+0 =18	18+50 =68	18+50 =68	18+50 =68	18+50 =68	63·0
	4	24+0 =24	24+0 =24	24+52 =76	24+52 =76	24+52 =76	44·8
	5	40+0 =40	40+0 =40	40+0 =40	40+54 =94	40+54 =94	45·4
	6	44+0 =44	44+0 =44	44+0 =44	44+0 =44	44+56 =100	46·8

By way of explanation, examine the row with a wall 4 m high. Irrespective of the height of the highest tide there will be an expenditure of £24 000 to build the wall. If the highest tide is less than 4 m there will be no flooding and no repair costs, and the total cost will therefore be £24 000. If, however, the highest tide exceeds 4 m as in the later columns of the table, then in addition to the £24 000 construction cost there will be a £52 000 repair cost giving a total cost of £76 000. Each location in the tabulation is completed in the same way. Each of the figures for total cost is in fact the pay-off associated with a particular height of wall if a particular high tide should occur. We can then associate this with the probability of different heights of tide and arrive at an expected value of the total cost for each of the four alternative wall heights. The last column in the table shows the expected value for each wall height and indicates this has a minimum value when the wall height is 4 m.

For most purposes this calculation is quite adequate, since the expected value is a reasonable concept on which to base decisions when each decision is one of many and there is a reasonable prospect that in the long run the expected value would be held to apply. There are, however, situations where the expected-value concept is not a very good one, and the following simple example will illustrate this point.

Risk

The calculation of expected value can be misleading because it takes little account of the importance of the risk to the person making the decision. It has already been shown that the expected value of the outcome of a series of gambling decisions based on the spin of a coin will be zero. This would imply that the decision-maker might be thought to be indifferent as to whether or not he involves himself in such a game. He may indeed be indifferent if the sum at stake is a few pence, or perhaps even £1, but he certainly would not be indifferent if the sum at stake was £1000 or more. The expected value in each case of a simple 50/50 game of chance such as spinning a coin is always zero, and therefore does not reflect at all the risks involved.

Table 8.4 illustrates this in a slightly different way where three alternative decisions are open to the decision-maker, each depending upon a 50/50 chance, but with different pay-offs in each case.

Table 8.4   Expected value of decision pay-offs

Decision number	0·5 probability of a win of	0·5 probability of a loss of	Expected value
1	+550	−50	+250
2	+3 600	−3 000	+300
3	+101 000	−100 000	+500

The expected-value criterion shows that decision 1 is the least attractive and decision 3 is the most attractive, but an examination of the win or loss figures would cast some doubt upon this. It is quite likely that most people would be prepared to put £50 at risk if there was an equal chance of a win of £550, that is, decision 1. Relatively few people would be prepared to enter into decision 2 which would put at risk £3000 with an equal chance of a win of £3600. It is highly improbable that anyone would opt for decision 3, which, while having an expected value of £500, puts £100 000 at risk in order to have a chance of a win of £101 000.

This demonstrates that simply to look at expected value is not adequate and a check should be made to see what the sums at risk are. Doing this in the case of the coffer-dam and looking back to table 8.3 it can be seen that the maximum outlay that could arise with each wall height, would be £68 000, £76 000, £94 000 or £100 000, with wall heights of 3, 4, 5 or 6 m respectively. The wall height that has the lowest absolute potential outlay is, as expected, the 3 m wall, but it does carry a very high risk, that is, a probability of 0·9 that this full expenditure will arise. This on its own is therefore not a very satisfactory criterion and it is useful to assess the risk associated with the wall height that gave the lowest expected value, that is, the 4 m wall with a maximum total outlay of £76 000. Both of these approaches look at the problem before the chance event, that is, the highest flood, has occurred. While it is obvious that the wall height must be determined before the flood occurs, it is none the less possible to look at the alternative decisions in retrospect after the event has occurred. It is necessary to look at each flood height in turn and ask the question 'If this flood height occurred, what would be the best decision, and by how much would each of the other decisions be worse than the best decision?'. This can be thought of as being a measure of the degree of *regret* at having made the wrong decision in the light of the particular height of tide that has occurred. The procedure for calculating this is set out in table 8.5, where entries are made by means of the following calculation.

Consider the first flood height, that is, less than 3 m. The best decision for the case where no flood exceeds 3 m, would obviously be to build the 3 m wall, and in this case there would be no regret since the best decision would have been made. If, however, the 4 m wall were built the sum expended would have been £24 000 compared with only £18 000 that was necessary for the 3 m wall. The decision would therefore be regretted to the extent of £6000, that is, £24 000 − £18 000. The figures in this column are completed by deducting the lowest figure in the column, that is, the £18 000 cost of the 3 m wall from each of the other costs. Proceeding similarly with the next column, the case where the maximum flood is between 3 m and 4 m, it is found that the best wall height to choose is 4 m. If the wall were only 3 m high the total expenditure involved would be £68 000 since it would be subject to flooding, compared with only £24 000 needed to build the 4 m wall. This decision would therefore be regretted to the extent of £44 000, the difference between £68 000 and £24 000. Again the column is completed by deducting the cost of the best solution for that flood height from all other cost figures. This type of calculation produces a complete regret matrix as in table 8.5.

Table 8.5 Regret matrix for coffer-dam

Maximum flood within the range (m)		up to 3	3–4	4–5	5–6	over 6
Probability that maximum flood is within this range		0·1	0·5	0·3	0·05	0·05
Wall height (m)	3	0	44	28	24	0
with corresponding	4	6	0	36	32	8
'regret' values (£ × 10³)	5	22	16	0	50	26
	6	26	20	4	0	32

Searching the table to find the decision that gives the lowest regret, no matter what the state of tide achieved, shows that the 6 m wall has a maximum value of regret of £32 000 whereas the 3 m, 4 m and 5 m walls have maxima of £44 000, £36 000 and £50 000 respectively. This analysis would therefore lead to the conservative view of building the 6 m wall and being safe.

There is, however, a problem with this method of calculating a regret matrix in that the method is not consistent. If, for example, the 4 m wall was for some reason not a possible decision, then the row in the table associated with the 4 m wall would not exist. If the calculation were then carried out to establish a new regret matrix it would appear as in table 8.6.

Table 8.6 Regret matrix for coffer-dam—omitting 4 m wall

Maximum flood within the range (m)		up to 3	3–4	4–5	5–6	over 6
Probability that maximum flood is within this range		0·1	0·5	0·3	0·05	0·05
Wall height (m)	3	0	28	28	24	0
with corresponding	5	22	0	0	50	26
'regret' values (£ × 10³)	6	26	4	4	0	32

This shows that the decision that has the lowest maximum value of regret is the 3 m wall. It is somewhat surprising that the omission of the 4 m wall which did not offer the lowest regret figure should now cause the decision based on the lowest absolute regret to change from the 6 m wall to the 3 m wall. This is partly due to the fact that these calculations of regret do not take account of the probability of any particular outcome arising. It is possible to make a calculation of expected value of regret in the same way that it is possible to calculate the expected value of cost and this would again indicate the best wall height to be 4 m.

One thing that becomes apparent in these various analyses is that the viewpoint of the decision-maker can affect the nature of the decision. It has been shown that if the decision-maker can afford to take the long-term view and accept losses as well as gains then it will probably pay him to make his decisions on the basis of expected value. If, however, the decision-maker is in the position that he has extremely restricted funds available, he may have to take the decision that has the lowest possible outlay irrespective of the conditions that arise. This leads to the decision in favour of the 3 m wall that can fairly clearly be seen to be a very short-term view since there is a high probability of flood arising. By comparison the regret-matrix concept leads to the fairly conservative decision to build the 6 m wall. Different methods of analysis therefore indicate different decisions depending upon the viewpoint of the decision-maker. This leads to the thought that it may be possible to take account of the position of the decision-maker and carry out an analysis on this basis, an approach that gives rise to the concept known as utility theory. This is designed to take account of the position of the decision-maker by assessing some measure of the importance of a gain or loss to him. For example, it would not be serious for an individual to be in a position where he may gain or lose £1, it would be more serious for him to be in a position of gaining or losing £100, extremely serious to be in a position of gaining or losing £1000, and possibly disastrous to be in a position of gaining or losing £10 000. This idea is pursued in the following paragraphs.

Utility Theory

Suppose you are in the position of having a bet of £1000 on the result of the spin of a coin. Would you be prepared as an individual to accept such a bet? If not, how much would you be prepared to pay to buy yourself out of this situation? It might be a figure such as £300. Secondly you are perhaps in a position where the probability of a loss is 0·9 and of a win 0·1; how much would you be prepared to pay to buy yourself out of the situation? You may well be prepared to pay £900 since the probability of a loss is very high. A further situation arises where the chance of a win is now 3 to 1 in your favour, that is, a probability of success of 0·75. You might be prepared to accept this, but might also be prepared to sell out your interest in such a bet for, say, £250. At what probability would you accept a bet and neither sell out your interest nor want to buy yourself out? It is likely that odds of 2 to 1 in your favour would allow you to accept the situation. The sums quoted here must obviously be the subject of agreement between buyer and seller, and can be thought of as insurance premiums. By their nature they must be slightly less advantageous than the true expected value, since someone else has to be rewarded for carrying the risk.

For the purpose of this analysis we arbitrarily assign to the gain of £1000 a utility of 1·0 since this is the maximum gain, and we assign to a loss of £1000 a utility of 0 since presumably you have zero utility for such a loss. In the same way that we calculated expected value we can calculate expected utility, that is, we multiply in any given situation the probability of a win with its utility value and then add up all such products. Hence in the first situation quoted above the expected utility would be $0·5 \times 1·0 + 0·5 \times 0 = 0·5$. This is equivalent to a loss of £300 since this is the figure you are prepared to pay to buy yourself out of the situation. We can therefore say that for you, the expected utility of 0·5 is equivalent to a loss of £300. We may similarly calculate the expected utility of the other situations quoted in the paragraph above, as in table 8.7.

Table 8.7   Expected utility

Situation number	Expected utility	Equivalent pay-off
1	$0·5 \times 1·0 + 0·5 \times 0·0 = 0·5$	$-£300$
2	$0·1 \times 1·0 + 0·9 \times 0·0 = 0·1$	$-£900$
3	$0·75 \times 1·0 + 0·25 \times 0·0 = 0·75$	$+£250$
4	$0·67 \times 1·0 + 0·33 \times 0·0 = 0·67$	£0

It is possible to plot a graph of expected utility value against equivalent pay-off and this is shown in figure 8.1. If the decision-maker is being reasonably consistent in his views of being more or less conservative, his pay-off assessments related to expected utility values should all fall on a reasonable curve. A more convex curve indicates a more conservative attitude, the straight line indicated on the diagram would relate to the straightforward expected-value concept, taking no account of the viewpoint of the decision-maker. Utility theory will not determine the shape of the line since this is affected by the decision-maker's general view and whether or not he is taking a conservative attitude to decision-making. What we can get from decision theory, however, is that we may recognise the basis on which decisions are made in this way, and can act consistently on that basis. It enables us to transpose the intuitive criteria used in making decisions on the spinning of a coin, to the more complex type of decision such as the coffer-dam construction already considered. It would strictly be necessary to construct a utility curve for a company making the decision, in much the

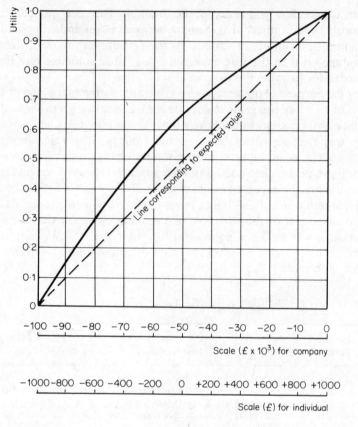

Figure 8.1   Personal and company utility functions

same way that the diagram was determined for the individual deciding on the results of spinning a coin. For the purpose of the example here, a different scale has been drawn in figure 8.1 to reflect the utility of a company for money. It is now possible to refer back to the coffer-dam example and another tabulation, table 8.8, is drawn up relating each possible cost level to its respective utility value.

The first column in the table lists in ascending order all the possible cost outcomes as listed in table 8.3. The second column gives the corresponding utility values taken from figure 8.1. Considering each possible wall height and referring back to table 8.3, it can be seen that there is a 0·1 probability of costs being only £18 000, and 0·9 probability of the cost being £68 000. Therefore enter 0·1 and 0·9 in the rows corresponding to £18 000 and £68 000 respectively, and multiply first 0·1 by 0·89, (the corresponding utility), and 0·9 by 0·44 (the corresponding utility). This procedure is carried out for each possible wall height and then at the foot of the table the sum

Table 8.8 Expected utility of coffer-dam decisions

Possible cost figures ($£ \times 10^3$)	Utility U	Wall height (m)							
		3		4		5		6	
		p	pU	p	pU	p	pU	p	pU
18	0·89	0·1	0·089						
24	0·85			0·6	0·510				
40	0·74					0·9	0·668		
44	0·70							0·95	0·664
68	0·44	0·9	0·396						
76	0·37			0·4	0·148				
94	0·12					0·1	0·012		
100	0·0							0·05	0·0
	Sum		0·485		0·658		0·680		0·664

expected utility value is calculated. In the example it can be seen that this has a maximum value when the 5 m wall is built. This would indicate that the 5 m wall offers the best height for the wall when account is taken of the decision-maker's point of view.

Summary

It is seen that there is no unique method of making the best decision. The approach will depend upon the decision-maker's general view—whether he is conservative or otherwise. All that can be gained from decision theory is recognition of the basis upon which decisions are made, thereafter enabling consistent action. It permits a test of whether intuitive decisions stand up to analysis, since although intuition may often give the right answer, it can sometimes neglect important factors. It is unreliable to attempt to solve complex problems by intuition alone, and therefore anything that can be done to back it up with reasoned analysis is beneficial. Decision theory therefore provides a rational and quantitative method for the examination of decisions. It has been found throughout industry that most decisions are not made in this way, and much research has been undertaken into the decision-making process. This opens up the subject into the areas of business strategy[2] and behavioural theory,[3] which try to assess the ways in which decisions are intuitively made.

References

1. C. W. Churchman, R. L. Ackoff and E. L. Arnoff, *Introduction to Operations Research*, Wiley, New York, 1957.
2. H. I. Ansoff, *Business Strategy*, Penguin, London, 1969.
3. R. M. Cyert and J. G. March, *A Behavioural Theory of the Firm*, Prentice-Hall, Englewood Cliffs, N. J., 1963.

9 Multi-stage Decisions Under Conditions of Uncertainty

It is often the case that a decision has to be taken at some time before the outcome of a future event is known, and it may arise that a whole series of decisions and chance events will present itself to the manager in such a way that he will feel somewhat bewildered by the range of inter-related problems confronting him. Single decisions can readily be assessed by the methods outlined in chapter 8, but where a series of decisions at separate times is encountered then a different technique is called for. The so-called decision tree[1,2] provides this method, and can be set out as shown below, illustrating an approach to a simple problem.

Decision Trees

Consider a contract that calls for bulk excavation work valued at £15 000 to be undertaken during winter months. The main contractor estimates that given good weather he can carry out the work himself at a cost of £10 000, but if the weather was bad it could cost £20 000. He knows that a sub-contractor will complete the work for a price of £12 000 irrespective of the weather. Assuming in this case that other factors such as quality of work and timing can be ignored, what decision should the contractor make? The decision-tree procedure is to write down first the decision point as a circle (figure 9.1), with the two alternative courses of action, namely to undertake

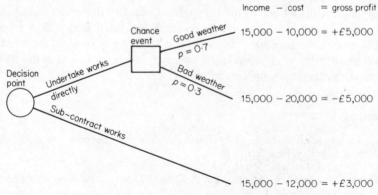

Figure 9.1   Decision tree on sub-contracting

the work directly or to sub-contract the work. If the work is done directly its cost will depend upon the chance of good or bad weather; in this example these are put down as simple alternatives. It is necessary to assess probabilities for these alternatives, and this can be based on recorded weather data for the area; in this case the relative probabilities are taken as $p = 0 \cdot 7$ for good weather and $p = 0 \cdot 3$ for bad weather.

The net cash result or pay-off for each possible outcome is calculated in a column at the right-hand side of the decision tree (note that a loss is shown as a negative profit). It is now possible to work backwards from these figures and calculate the expected value of profit for the two alternative decisions which can be made. The concept of expected value (e.v.) and its use are described in chapter 8.

e.v. of gross profit if work done directly

$$= 0 \cdot 7 \times 5000 + 0 \cdot 3 \, (-5000)$$
$$= 3500 - 1500 = £2000$$

e.v. of gross profit if work sub-contracted

$$= 1 \cdot 0 \times 3000 = +£3000$$

On this basis therefore it would appear to be preferable to sub-contract the work.

The problem described above is only a single-decision problem and could have been handled by simply calculating an expected value directly. However, the method can easily be extended to cover a more complex problem involving a series of decisions, as described in the following problem concerning a contractor who is faced with having to make a decision on whether or not to make a capital purchase, prior to knowing whether or not the purchase is essential. The particular case is as follows.

Quarry Materials Supply

The contractor knows that there are two major road-contracts coming up in his area within the next few years. The first of these, X, is coming out to tender shortly, but contract Y will not come out to tender until contract X is completed. It is already known that contract X will require fill material in large quantities and this will have to be supplied to a very tight specification; it is thought that the only reasonable way of complying with the specification is to supply material from one particular existing quarry. Contract Y which will follow some time later is of a slightly different nature and will be able to use fill material from a number of different sources including the quarry that is to supply contract X.

The contractor has learned that the particular quarry is being put up for sale, and he has to decide whether or not he should attempt to negotiate purchase of the quarry. It is fairly certain that if he does not buy the quarry one of his competitors might attempt to do so, and should he then be successful in obtaining contract X he will either have to buy the quarry from his competitor at an increased price, or buy material from the quarry at a relatively high price. The decision has to be made before the result of the tender for contract X is known and before anything much is known about contract Y.

The contractor decides that he is not prepared to tender for job X unless he has bought the quarry, since use of its material is virtually essential, and he is not prepared to take the risk that one of his competitors might force him to pay an excessively high price for material from the quarry. He therefore decides that if he does not buy the quarry he should not bid for job X, but he might subsequently bid for job Y where there are several sources of material available. The contractor wishes to make a logical analysis of the decisions facing him and he therefore writes down a number of decisions that he has to make as follows.

(1) Buy the quarry and bid for job X, or forget both?

(2) If he buys the quarry and wins job X does he keep or sell the quarry prior to bidding for job Y?

(3) If he buys the quarry and loses job X, does he keep or sell the quarry prior to bidding for job Y?

(4) If he does not bid for job X does he attempt to buy the quarry prior to bidding for job Y?

In these it is assumed that once contract Y has been let the quarry is not available to the contractor, and it is further assumed that once both X and Y contracts are completed, the quarry still has some residual value. It is possible to set down the alternative courses of action in a decision-tree diagram using the following notation.

A decision point is shown as a circle with a number of alternative courses of action, any one of which may be taken; for example, to buy the quarry or not to buy the quarry. A square box is used to indicate a chance event that has a number of possible alternative outcomes; for example, winning or losing a particular contract. Since these are chance alternatives it is usual to assign to each of them a probability that it will arise, remembering that the sum of the probabilities of the alternative outcomes must be unity. The full range of alternative decisions and chance events is set out in the tree in figure 9.2.

It is a simple matter to follow through the logic of the decision tree, for example, if the contractor decides to buy the quarry he is then following the path 1 to A on the diagram. If he is then successful in winning contract X he then follows the path from A to 2. At decision point 2 he has to make a positive decision of whether to keep or sell the quarry and if he should decide at that point to sell this is indicated by the line from 2 to C. At chance event C he will either succeed or fail in his attempt to win contract Y and follow the appropriate path. Other paths indicate the alternative choices and alternative chances or success or failure in bidding for contracts. This diagram only sets out the logic of the case and it is necessary, in order to carry out some evaluation, to assign numerical values to the following.

(1) Estimates of the purchase price of the quarry at the outset and/or at later times.

(2) The resale of the quarry taking account of its depletion in value as material is extracted for use in contracts X and Y.

(3) The profit that may be made on each of contracts X and Y both with and without ownership of the quarry.

(4) The probabilities of winning or losing each of the contracts X and Y.

All of these numerical values are now entered into the diagram, figure 9.3, showing cash outflow as a negative figure and cash inflow, that is, profit, as a positive figure.

The first step in the calculation is to assess the net cash-flow that would result from each possible sequence of decisions and chance events, for example, the sequence from 1 to A to 2 to B has a net outcome calculated as follows.

Buy quarry (purchase)		− 1000
Profit from job X	+400	
Keep quarry (interest, etc.)		− 50
Profit from job Y	+300	
Ultimate sale of quarry net outcome	+700	

$$1400 - 1050 = 350$$

This figure is calculated for each alternative outcome and entered in the last column of figure 9.3.

The next step in the procedure is rather like that adopted in dynamic programming, which is discussed later. The procedure is to take the latest decision first and say 'If I were in this position, what would be the best decision for me to take?'. In the example under consideration this would mean that the contractor is at decision point 2 having bought the quarry and successfully completed contract X at a profit of $+400$. His decision then is whether or not to sell the quarry and the procedure is to calculate the expected value of the two alternative outcomes. The expected value of cash flow following chance event B can be calculated as follows.

$$\text{e.v.}_{\text{B}} = 0 \cdot 3 \times (300 + 700) + 0 \cdot 7 \times (0 + 800) = +860$$

For the contractor to keep the quarry there is a cost of 50, that is, a cash flow of -50 and therefore the expected value of making the decision to keep

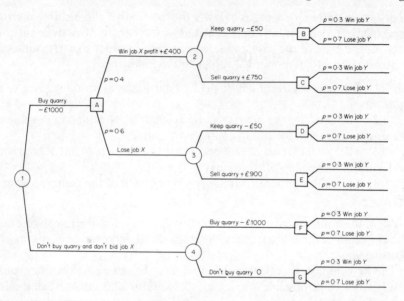

Figure 9.2   Decision tree on contract bidding and quarry purchase

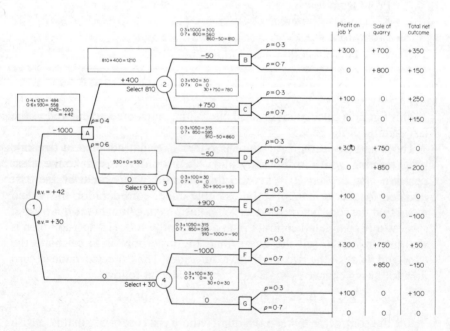

Figure 9.3   Decision tree—pay-offs, probabilities and expected values

the quarry at decision point 2 is $860 - 50 = 810$. The expected value of the chance event C can similarly be calculated as follows.

$$\text{e.v.}_C = 0\cdot3 \times (100+0) + 0\cdot7 \times (0+0) = +30$$

The direct cash-flow resulting from selling the quarry at decision point 2 is an income of 750 and therefore the expected value of the decision to sell the quarry at decision point 2 is $750 + 30 = 780$. The two alternative outcomes of decision 2 can therefore be compared, namely 810 against 780 and it can be concluded that if decision point 2 is arrived at, then the optimal decision at that point is to keep the quarry, with an expected value of the outcome of $+810$. Similar calculations are shown below to assess the expected value of chance events, D, E, F and G.

$$\text{e.v.}_D = 0\cdot3(300+750) + 0\cdot7(0+850) = 910$$
$$\text{e.v.}_E = 0\cdot3(100+0) + 0\cdot7(0+0) = 30$$

Then compare

$$910 - 50 = +860 \text{ for decision 3 to D}$$

with

$$30 + 900 = +930 \text{ for decision 3 to E}$$

Hence the optimal decision at 3 is to sell the quarry with an expected value of the outcome of $+930$.

$$\text{e.v.}_F = 0\cdot3(300+750) + 0\cdot7(0+850) = 910$$

$$\text{e.v.}_G = 0\cdot3(100+0) + 0\cdot7(0+0) = 30$$

Then compare

$$910 - 1000 = -90 \text{ for decision 4 to F}$$

with

$$30 + 0 = +30 \text{ for decision 4 to G}$$

Hence the optimal decision at 4 is not to buy the quarry.

Having now calculated the expected value of the optimal decisions at decision points 2, 3 and 4 we move back in time and consider chance event A evaluated as follows.

$$\text{e.v.}_A = 0\cdot4(400+810) + 0\cdot6(0+930) = 1042$$

Then compare

$$1042 - 1000 = 42 \text{ for decision 1 to A}$$

with

$$30 + 0 = 30 \text{ for decision 1 to B}$$

Hence the optimal decision at 1 is to buy the quarry and bid for contract X.

The conclusions that may be drawn from this analysis, are therefore as follows.

(1) It is only just advantageous in the first place to buy the quarry.

(2) Having won job X it is preferable to keep the quarry.

(3) If job X is lost, it is preferable to sell the quarry.

(4) If job X is not tendered for, then it is preferable not to buy the quarry prior to bidding for job Y.

There are again many factors that may have to be taken into account in making these decisions, but at least an idea of the possible financial outcome of the decisions can be evaluated and used as an aid to judgement and decision-making. As in the previous example, the decisions may well be sensitive to some of the assumed data, and in working situations it would be advisable to test the sensitivity of decisions to the accuracy of input data.

Decision trees may be used in situations where a number of decisions and chance events interact and make it difficult to assess the full implications of decisions at the time they are made. They are widely used in development projects when decisions have to be made about continuing or stopping particular lines of work.

Dynamic Programming

As in the case of linear programming described in chapter 5 the word 'programming' means mathematical manipulation and not planning. The essential feature of dynamic programming is that it can cope with multi-stage decisions by taking account of future decisions yet to be made or future events yet to happen. The decision-tree method described earlier in this chapter is one form of dynamic programming, but there are others that may be used in different situations. While the method may seem complex the principle is illustrated by the following numerical problem. Figure 9.4 is a number matrix with an array of values in large type; the procedure is to select a number from the first column then move to a number in the second column immediately above *or* adjacent to *or* immediately below the first number selected. A move is then made to the third column one place up or directly across or one place down, and similarly column by column right through the matrix. The objective is to find the path that maximises the sum of values passed through in moving from the first to last column. It would of course be possible to enumerate all possible paths, but in the example quoted there would be nearly two thousand paths to calculate.

The procedure is to start in the second column from the right, e, consider each box in turn down the column, and ask the question 'If a path arrived in this box from the left, which would be the best way from here?'. For example taking the first number in column e, value 4, the maximum value thereafter is given by moving to the adjacent box value 3, giving a total for the part of the

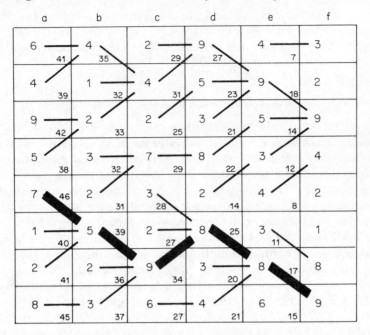

Figure 9.4   Number matrix—dynamic programming

path *ef* of $4 + 3 = 7$. Similarly taking the second box in column e, value 9, the maximum value would be found by moving one row down in column f, giving a value for *ef* of $9 + 9 = 18$. These cumulative values are entered in small type in the bottom right-hand corner of each box, and a line marked in to show the route which achieved it. When this has been done for all boxes in column e, the next step is to repeat the procedure for column d. Taking for example the third box in column d it can be seen that the best path thereafter is to move up one row in column e giving a maximum cumulative value for *def* from that box in d of $3 + 18 = 21$. Again this cumulative value is entered in small type in the bottom right-hand corner of the box, and a line marked in to show the route. If this calculation is carried out for each box in column d the lines marked indicate the answer to the question 'If a path arrived in this box from the left, which would be the best way from here?'.

Continuing this process of calculation moving in turn through columns d, c, b and a will finally yield the path through the matrix that maximises the sum of the values, that is, the path which satisfies the objective of the problem. It is seen to be $7 + 5 + 9 + 8 + 8 + 9 = 46$. This calculation has involved about one hundred additions of a pair of numbers—a much smaller task than the summation of nearly two thousand sets of six numbers needed to evaluate all possible combinations.

In this simple numerical example the problem is deterministic, and there is no element of uncertainty. This chapter is however concerned with uncertainty, and the method may be used under such conditions. Again the procedure is illustrated by reference to an example.

Marine Works—Method Decision by Dynamic Programming

A contractor is planning the construction of a cooling-water intake structure that consists of a shallow pontoon with adjustable legs. The structure is to be assembled on shore in two stages and floated out on a spring tide. It is anticipated that the work will take three months to complete.

The contractor is faced with a number of decisions, which may be summarised as follows. He can construct a coffer-dam, at a cost of £5000, that will ensure that work can continue uninterrupted during the first spring tide. It will, however, take time to construct, and it thereby reduces his chance of being up to programme by that point. He assesses that there is only a 0·5 probability of being up to programme, meaning that there is a corresponding 0·5 probability that he will run late and incur additional costs of £10 000 in catching up before the second spring tide. It is possible, however, to dispense with the coffer-dam to gain more time, and reduce the probability of being behind programme at the first spring tide to 0·33. However, the consequences of being late and with no coffer-dam are more serious since damage to the temporary works would arise, leading to a repair cost of £14 000 in addition to the £10 000 necessary to catch up and complete stage one before the second spring tide. The contract calls for completion and launching on the third spring tide, but if it were possible to complete and launch by the second spring tide, a bonus of £15 000 would be payable. In order to gain the bonus it would be necessary to overlap the completion of stage one with the stage-two work at an additional cost of £12 000. The nature of the stage-two work is such that it could be spread over two months at a saving of £11 000 provided that stage-one work was completed by the first spring tide. The decisions facing the contractor are therefore

(1) Should he build the coffer-dam, or risk flooding?

(2) If stage one is complete by the first spring tide should he reduce production rate and spread stage two over two months?

(3) Should he expend money in an attempt to gain the bonus?

The alternative courses of action and outcomes are set out in figure 9.5, and the various costs are as follows.

Stage-one work	−£100 000
Stage-two work	−£120 000
Repairs if works are flooded	−£14 000
Build coffer-dam	−£5 000

Catching-up cost if stage-one is late	−£10 000
Acceleration cost to complete stage-two early	−£12 000
Bonus for early completion	+£15 000
Saving from spreading out stage-two	+£11 000
Contract valuation	+£250 000

Figure 9.6 has the same shape as figure 9.5 but now includes the costs in £ × 10³ of the alternative paths. Note that the nominal cost of stages one and two and the contract valuation are omitted from this calculation, since they are common to all alternatives. Costs are shown as negative sums, savings and bonus payments as positive sums.

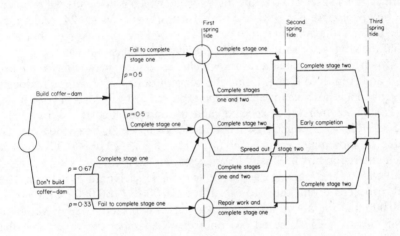

Figure 9.5 Marine works—alternative decisions

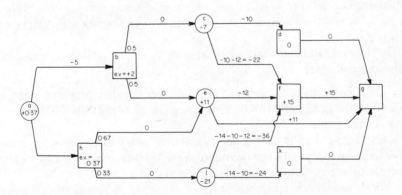

Figure 9.6 Marine works—costs of alternative decisions

The procedure for calculation is similar to that outlined in the number matrix earlier in this chapter. In this case there is only one possible path from each of d, f, k (at the second spring tide) and therefore no choices have to be made at that point. It is, however, necessary to enter into each of these boxes the cumulative cost position. At the first spring tide choices are available at c, e and j. At box c it is necessary to compare $cd = -10 + 0 = -10$ with $cf = -22 + 15 = -7$. Since the latter has a higher absolute value it is preferred and entered as the cumulative cost in box c. This means that if the contractor found himself in the position of box c (having built the coffer-dam and failed to complete stage one on time), he should then accelerate his work and complete it ahead of schedule. At box e compare $ef = +15 - 12 = +3$ with $eg = +11$; select eg as the higher value which is now entered in box e. At box j compare $jf = +15 - 36 = -21$ with $jk = 0 - 24 = -24$; select jf as the higher value and enter into box j. These selections have now indicated what the contractor should do if he found himself in situations c, e, j.

Moving left one stage there is now a chance element in moving from b to c or e and from h to e or j. Using the concept of *expected value* previously described it is possible to calculate the expected value of costs beyond b and h. Taking box b first the expected value of subsequent costs is $0.5 \times (-7) + 0.5 \times (+11) = +2$, which is then entered in box b. Taking box h the expected value of subsequent costs is $0.67 \times (+11) + 0.33(-21) = +0.37$ which is then entered in box h. Finally it is necessary to compare $ab = +2 - 5 = -3$ with $ah = 0 + 0.37$, giving a higher value to ah which is entered in box a. The net proceeds from the project are £250 000 − £100 000 − £120 000 = £30 000 and the above analysis has given an expected value of variation from this of +£370, based on route ah. Remember, however, that the *expected value* is a generalised concept and is not a particular outcome that can arise. If route ah is chosen then it will either follow $aheg$ with a value of +11 or it will follow $ahjfg$ with a value of −21. It may in practice not be acceptable to risk an additional cost of this magnitude and the alternative initial decision ab may be preferred. Although it has a lower expected value its actual alternatives are $abcfg$ with a value of −12 or $abeg$ with a value of +6.

The decisions indicated by the above analysis, corresponding to the initial questions asked are

(1) It is slightly preferable not to build the coffer-dam provided that the risk of a cost of £21 000 can be run. Since the gross profit is £30 000 this appears acceptable.

(2) If stage one is completed by the first spring tide it appears to be advantageous to spread stage two out over the two remaining months, rather than to try to complete one month early and gain a bonus.

(3) It is only advantageous to spend money in an attempt to earn the bonus if stage one is not complete by the first spring tide, irrespective of whether the coffer-dam has been built or not.

The method of dynamic programming described in the above example is essentially the same as in the earlier numerical problem. The procedure is to start at the end of the project and work backwards to the beginning, asking at each stage 'If this situation should arise, what is the best way forward from here?'. It is of course essential to formulate any problem in the way set out here, and assessments have to be made on probabilities of chance events, and estimates made of the costs or savings that may arise from various courses of action. Difficulty in making such estimates may appear to invalidate the method, and in the example chosen here it appears that the outcomes of the project would not be very different whichever decision is taken. There are many occasions, however, when the alternatives do not give such close values, and it is clear which is the best series of decisions. In these cases it is valuable to test the sensitivity of the decisions to changes in the estimates.

In the marine example just described it can be shown that the advantage of saving time by not building the coffer-dam would be eliminated if the probability of completing stage one in time fell from 0·67 to 0·56. Similarly it can easily be seen that the bonus would not be worth striving for if it had a value less than £12 000. Again it must be emphasised that it would be dangerous to rely exclusively and explicitly on the method shown here, since it necessarily involves some simplification. It does, however, provide a rational means of comparing alternatives, and it may then be possible to consider these in perspective when other non-quantifiable factors are being considered. This problem is essentially a simple one that could have been solved by common sense. In reality problems are usually much more complex, and not amenable to simple arithmetic; it is in such situations that the more formalised approach of dynamic programming is of advantage.

References

1. R. A. Johnson, W. T. Newell and R. C. Vergin, *Production and Operations Management, A Systems Concept*, Houghton Mifflin, Boston, Mass., 1974.
2. V. C. Hare Jr, *Systems Analysis, A Diagnostic Approach*, Harcourt Brace and World, New York, 1967.

10 Competitive Bidding

This chapter will primarily be of interest to contractors who are interested in improving their estimating and tendering practice. An understanding of the quantitative aspects of bidding will also be of interest to consulting engineers and others responsible for the placing of contracts. It can therefore be of help both to contractors and to clients that the interactions of bid levels, success rates and number of tenderers should be well understood.

To many people the whole subject of bidding and tendering appears to defy analysis and is cloaked in a certain amount of mystery. One reason for this is that there are so many variables that are not well understood in themselves and certainly not understood when they interact, that it becomes very difficult to try to make specific predictions. Some of the more important of these variables are related to the cost of completion of the work, for example the price of materials, labour rates, labour productivity, plant usage, ground conditions, weather, variations instructed as to the detailed work to be completed, and additional costs associated with delays caused by shortage of materials or lack of information. A second reason for the mystery surrounding bidding is that it is a very sensitive area to many contractors and they are unwilling to discuss it. This is largely due to the desire to keep methods of estimating work and the prices used confidential to the company, but a more subtle and seldom-quoted reason for secrecy is that many contractors perhaps do not themselves fully appreciate how their prices are arrived at, and may indeed have very little or no systematic approach to bidding whatsoever.

Clients' Objectives

Before making a quantitative analysis of bidding strategy it is worth discussing the objectives of both the contractor and the client in a bidding situation. The client's objective is to enter into a contract that will ensure the completion of the work within the required time period at the lowest price consistent with an acceptable quality of workmanship. It is common but not universal practice to accept the lowest tender submitted. There should in practice be very few occasions on which the second lowest bid is thought to be preferable to the lowest bid unless their tendered prices are extremely close. It is sometimes argued that the lowest tender may not be always acceptable since there could be considerations other than price to take into account, for example the lowest bid might be submitted by a contractor with a reputation for indifferent work. If this is the case then it is not right ever to consider a bid from such a contractor and he should not be asked to submit a tender in the first place. Tenders should only be sought and accepted from contractors who are thought to be capable of satisfactory completion of the contract in all its aspects. It is argued by some writers on bidding that the lowest tender should never be accepted and that the contract should always be awarded to the second lowest bid. This is based on the argument that it is likely that the contractor submitting the lowest bid will have underestimated the cost of doing the work and will consequently be taking on the contract at a loss.

While the profitability of the contract may be of no immediate concern to the client, it may well have a secondary effect since frequently when contractors realise they are in a loss-making situation on a contract many difficulties arise and the contract may go sour. While this argument may have some merit it is doubtful whether the principle is ever used in practice. It is, however, relevant to the question of the number of contractors who are invited or permitted to submit tenders for a particular contract. Clients may quite reasonably feel that if they increase the number of tenderers then they will increase the chance of receiving a low bid. This is undoubtedly true but it is also true that in so doing they will increase the chance of receiving a ridiculously low bid, and therefore one that will lead certainly to a loss for the contractor and possibly to great difficulty for the client himself. In addition if it is general practice to invite a large number of contractors to submit bids for every contract, then each contractor's average success rate will be low and he will have to carry a relatively high level of tendering costs. The cost of all tenders both successful and unsuccessful must be borne out of the revenue derived from successful tenders and therefore if it is general practice to call for a large number of tenders, then this will mean that all successful contracts will have to carry a substantial proportion of abortive tendering costs.

These two factors together suggest that it is not in the interest of clients to seek large numbers of bids in order to obtain low prices and indeed there is some evidence to suggest that from the client's point of view the optimum number of tenderers is about eight. This may in part be due to an attitude on the part of contractors who, when they realise that they are one of a large number submitting bids, do not make a very serious attempt at estimating for the work and put in relatively high bids; if this view happens to be taken by all contractors on this specific contract then the resulting successful tender will be a relatively high one.

Contractors' Objectives

The objectives of the contractor are a little more difficult to assess. The most commonly stated objective will be the maximisation of profit, but this in itself is not clear. First this might mean the maximisation of absolute profit in

£s per annum. It might mean the maximisation of profitability, that is, the profit as a percentage of turnover, or it might mean the maximisation of profit as a percentage of capital employed in the business. While these three interpretations may lead to similar policies they are not exactly the same and the management of a contracting company should be clear as to which of these three interpretations of profit it is seeking. This is essentially a financial policy decision and should be taken at the highest level of the company.

An alternative objective that is not consistent with the maximisation of profit, is that which seeks to increase the level of operation of the company, that is, to increase its turnover. Obviously the turnover will be maximised by submitting a large number of low bids thereby ensuring a high success rate, but of necessity at the same time accepting that many contracts will be undertaken at low profit or even at a loss. In the extreme case very low bidding would ensure one hundred per cent success but would at the same time ensure an overall loss that is not acceptable.

It would therefore appear that to seek a high turnover is not a valid objective for the company but it is interesting to note that it might be a reasonable objective for individuals employed within the company. It is fairly natural that an estimator employed within a contracting company will be pleased to win a substantial number of contracts, and indeed he will probably be thought to be a bad estimator if he has a low success rate. The subsequent satisfactory completion of a contract is not normally his concern and it may therefore be thought to be in his personal interest to keep tenders slightly low in order to enhance his success rate. Furthermore the opportunities for individual development and promotion within an organisation will increase if the total level of activity of that organisation grows, and again therefore it may be in the interests of individuals to try to increase the tendering success rate. There is therefore the possibility of a conflict arising between the objectives of the company and individuals employed in the company, and this should be recognised and account taken of it in tendering policy.

The above two major objectives of contractors may both apply and the resultant policy may be a compromise between them. In times of a booming economy it may make sense to maximise the profitability or return on capital employed but at times of lower economic activity when there are less contracts available the better policy may be to seek a turnover that will keep the plant and human resources of the contractor usefully employed and make a contribution to the maintenance of the company. In this case of compromise the objective of the company then is essentially that of survival. This is the conclusion reached by Ansoff in some of his writings on corporate strategy and business policy.[1,2]

From time to time there will be subsidiary objectives expressed by a contractor. He may have a particular desire to be successful in tendering for a contract in order to obtain experience of work of a particular type or in a new geographical area. He may also at times seek to keep his competitors out of a particular area or even to deprive them of work altogether. The methods in this chapter show first how to strive for the primary objectives but also indicate ways in which the secondary objectives may be taken into account.

Cost Estimate

Bidding is concerned with making decisions under conditions of uncertainty. The first uncertainty in submitting a bid relates to the cost of completing the work if the bid is successful, there being many areas of uncertainty including those listed in a previous paragraph. The second main area of uncertainty relates to the bids that will be submitted by competitors and it is important that these two areas of uncertainty should be considered separately.

An estimator should always give his best estimate of the *actual* cost of the completion of the work, if possible giving some idea of his confidence in that best estimate by quoting also upper and lower limits to the cost of completing the work. This cost estimate should take account of possible variations due to technical factors such as price changes, productivity, delays and so on, but should *exclude* consideration of how keen the company is to obtain the contract. This latter point is a matter of a business policy decision and the strategy adopted by the company should be decided at a high level. It cannot be over-emphasised that the estimator should always work on the principle that the bid will be successful and that his task is to produce a realistic assessment of what the cost of completion of the work will be. It is subsequently the responsibility of senior management in the company to decide on the bid price to be submitted taking account of the cost estimate, the objectives of the company, the general economic climate, and the workload situation of the company. If the estimator is allowed to build into his costs such factors as the company's keenness to win then his estimate will not be truly an estimate of the cost of the work and it will become correspondingly difficult to apply an appropriate mark-up in order to arrive at a tender price.

In order to study bidding strategy in a quantitative manner Friedman's simple approach is employed first of all as illustrated by the following example. Consider first the simplified case where a company knows that the contract under consideration will cost exactly £100 000 to complete. In doing this we are assuming that there is no variation in the contract cost but account will be taken of this complication at a later stage. It is also assumed that we are able to assess the probability of winning the contract with bids at various levels. By making this assumption we have overcome the problem of assessing what the probability of success actually is, but again consideration of this problem will follow later. In this example we will seek to maximise the expected value of profit on the contract, that is, the expected average level of profit that would be achieved by carrying out large numbers of contracts of this type. Remember that the expected value of profit is not an actual value that will ever occur but is the long-term average value taken over a large number of contracts. Readers who are uncertain about the concept of expected value should refer to chapter 8 where it is discussed more fully.

Table 10.1   Expected value of bids—Friedman

Bid (£)	Probability of success	Profit (£)	Expected value of profit (£)
95 000	1·00	−5 000	−5000
100 000	0·95	0	0
105 000	0·55	+5 000	+2750
110 000	0·36	+10 000	+3600
115 000	0·25	+15 000	+3750
120 000	0·17	+20 000	+3400
125 000	0·12	+25 000	+3000
130 000	0·08	+30 000	+2400

From the above it can be seen that the maximum expected value of profit is £3750 with a bid of £115 000. For simplicity bids at intervals of £5000 have been considered, but of course in practice it would be possible to make intermediate bids.

It is of interest to investigate how sensitive the bid decision is to accuracy in cost estimate, and to find the range of cost over which the bid of £115 000 is the best. Consider that our estimate of £100 000 is in error and could have a lowest value of C_L or a highest value of C_H. It will just pay to change our bid from £115 000 to £110 000 if the cost C_L is such that the expected value of profit is the same with either bid, namely

$$(110\,000 - C_L) \times 0.36 = (115\,000 - C_L) \times 0.25$$

This occurs when $C_L = 98\,636$.

Similarly it will just pay to change the bid from £115 000 to £120 000 if the cost C_H is such that the expected value of profit is the same with either bid, namely

$$(115\,000 - C_H) \times 0.25 = (120\,000 - C_H) \times 0.17$$

This occurs when $C_H = 104\,375$.

Hence the decision to bid £115 000 will give the maximum expected value of profit over the range of £98 636 to £104 375 for the cost of completing the work. This means that an exact cost estimate may not be vital to making the right bid decision, and some inaccuracy can be tolerated. This sensitivity test makes the assumption that probability of success is independent of estimate accuracy.

Probability of Success

In the above example it was assumed that the probability of success was known, but this knowledge would have to be gained by the study of a wide range of data. The probability of success in a particular contract can be assessed from previous tendering-practice of competitors. It is often possible to learn the bids submitted by competitors either directly from the client or from other sources. On the somewhat optimistic assumption that this information is available it is possible to study the bidding behaviour of a competitor in the following way.

Table 10.2   Competitor's bidding behaviour

Contract number	Our estimate of the cost of the work (£)	Bid submitted by competitor A (£)	Ratio of bid of A: our cost
1	100 000	114 000	1·14
2	60 000	72 000	1·20
3	242 000	261 000	1·07

Given data for a large number of contracts it is possible to prepare a frequency distribution table of the ratio 'bid of A:our cost' calculated as follows for one hundred contracts.

Table 10.3   Frequency of competitors bids

Ratio of bid of A:our cost	Number of times occurring	Probability of this ratio	Probability of this ratio or higher
0·91 to 1·00	1	0·01	1·00
1·01 to 1·10	5	0·05	0·99
1·11 to 1·20	14	0·14	0·94
1·21 to 1·30	28	0·28	0·80
1·31 to 1·40	35	0·35	0·52
1·41 to 1·50	14	0·14	0·17
over 1·50	3	0·03	0·03

The last column in this table is calculated by successively deducting the probability of a particular ratio occurring from the previous line; for example, there is a probability of 1·00 that all ratios will be above 0·91, but since there is a 0·01 probability of the ratio being in the range 0·91 to 1·00 then the probability of the ratio being greater than 1·00 is $(1·00-0·01) = 0·99$.

Expected Value of Bids

If we now consider that we are bidding against competitor A only we can assess our chance of success at various levels of bid as follows.

Table 10.4 Probability of beating a competitor

Ratio of our bid: cost	Probability (p_A) that our bid is lower than the bid of A
0·90	1·00
1·00	0·99
1·10	0·94
1·20	0·80
1·30	0·52
1·40	0·17
1·50	0·03

This will then enable us to calculate the expected value of profit if we compete in a series of contracts with competitor A only.

Table 10.5 Expected value of profits—single competitor

Ratio of our bid:cost	Profit:cost	Expected value of profit:cost
0·90	−0·10	−0·10 × 1·00 = 0·100
1·00	0	0 × 0·99 = 0
1·10	0·10	0·10 × 0·94 = 0·094
1·20	0·20	0·20 × 0·80 = 0·160
1·30	0·30	0·30 × 0·52 = 0·152
1·40	0·40	0·40 × 0·17 = 0·068
1·50	0·50	0·50 × 0·03 = 0·015

Table 10.5 shows that the maximum expected value of profit is 16 per cent with a bid:cost ratio of 1·20, that is, a 20 per cent mark-up on cost.

It is most unlikely that any contractor will ever be in the position of competing consistently against one competitor, and it is much more common that he will be competing against several. Suppose these are A, B, C, D; as described above for competitor A it is possible to study each competitor's bidding behaviour and assess the probability of being able to beat each competitor at each level of bid:cost ratio. The probability of being able to beat all competitors simultaneously in a tender is the product of the individual probabilities of beating each of them, and hence it is possible to tabulate the probability of winning a contract with various levels of bid:cost ratio.

Table 10.6   Expected value of profits—several competitors

Ratio of our bid:cost	Profit:cost	Expected value of profit:cost profit $\times p_A \times p_B \times p_C \times p_D$
0·90	−0·10	$-0\cdot10\times1\cdot00\times1\cdot00\times0\cdot98\times0\cdot97 = -0\cdot0951$
1·00	0	$0\times0\cdot99\times0\cdot98\times0\cdot94\times0\cdot96 \quad = 0$
1·10	0·10	$0\cdot10\times0\cdot94\times0\cdot93\times0\cdot92\times0\cdot95 = 0\cdot0764$
1·20	0·20	$0\cdot20\times0\cdot80\times0\cdot74\times0\cdot70\times0\cdot82 = 0\cdot0680.$
1·30	0·30	$0\cdot30\times0\cdot52\times0\cdot48\times0\cdot45\times0\cdot56 = 0\cdot0189$
1·40	0·40	$0\cdot40\times0\cdot17\times0\cdot15\times0\cdot15\times0\cdot20 = 0\cdot0003$
1·50	0·50	$0\cdot50\times0\cdot03\times0\cdot02\times0\cdot03\times0\cdot05 = 0\cdot0000$

Table 10.6 indicates that maximum expected value of profit will be given by a bid:cost ratio of 1·10.

The calculations above depend on knowing the past bidding-performance of individual competitors. Such information may be difficult to obtain, but it is common practice for clients to inform all tenderers of the full list of tender sums submitted, without identifying the competitor associated with each bid. If information is known in this form it is possible to treat all competitors as average, and to establish the frequency distribution of bids as before. If we are now faced with a decision in a tender where there are a number of competitors we simply calculate the expected value of profit as before, but in place of using $p_A \times p_B \times p_C \times p_D$ we use $(p_{av})^n$ where n is the number of competitors. In the table below the calculation allows for four *average* competitors.

Table 10.7   Expected value of profit—average competitors

Ratio of our bid: cost	Profit:cost	Expected value of profit: cost profit $\times (p_{av})^n$
0·90	−0·01	$-0\cdot10\times1\cdot00^4 = -0\cdot1000$
1·00	0	$0\times0\cdot97^4 = 0$
1·10	0·10	$0\cdot10\times0\cdot93^4 = 0\cdot748$
1·20	0·20	$0\cdot20\times0\cdot76^4 = 0\cdot0667$
1·30	0·30	$0\cdot30\times0\cdot50^4 = 0\cdot0188$
1·40	0·40	$0\cdot40\times0\cdot16^4 = 0\cdot0003$
1·50	0·50	$0\cdot50\times0\cdot03^4 = 0\cdot0000$

In the example given, the result using individual competitors is virtually the same as when all are treated as average competitors. It is therefore unlikely that it is worth the extra effort involved in the former calculation unless there are significant differences between the behaviour of different competitors, and it is known which of them are tendering for the contract under consideration.

Success Rate Related to Mark-up

The above analysis depends upon knowledge of the values of both winning and losing bids, and this information may not always be available. Another approach has therefore been developed by Gates[3] which depends upon knowing only our own estimate of the cost, our own bid and the value of the winning bid for each contract. The procedure is to examine as many past contracts for which we have tendered as possible and tabulate the information as follows in table 10.8.

Table 10.8   Comparison of winning and our bids—Gates

1	2	3	4	5	6	7	8
	Lowest competitive bid	Our bid	Ratio of lowest competitor to our bid	Bid difference	Our mark-up	Mark-up to win	Descending order of mark-up
1	120 400	123 000	0·9789	−0·0211	+0·0165	−0·046	8

Column 1   Contract identification

Column 2   Lowest competitive bid, that is, the winning bid, or in the case of contracts that we won, it is the second lowest bid

Column 3   Contains the value of our bid

Column 4   Is the ratio of lowest competitor to our bid

Column 5   Is the difference between the lowest competitive bid and our bid expressed as a proportion of our bid, that is, it is column 4 minus 1·00

Column 6   Is the mark-up we used in arriving at the bid expressed as a proportion of the bid

Column 7   Indicates the mark-up needed to win expressed as a proportion of our bid and is calculated by adding columns 5 and 6

Column 8   Ranks in descending order of magnitude the mark-up indicated in column 7

This calculation of mark-up is not quite consistent and the procedure outlined by Gates is not exactly as is given below, but amounts effectively to the same calculations and it is thought that the following method reduces the amount of arithmetic needed. The calculation is set out in table 10.9, each column of which contains the figures described.

Table 10.9   Winning bids and our cost

1 Lowest competitive bid	2 Our cost	3 Ratio low competitive bid:our cost	4 Mark-up needed to win (%)	5 Descending order of mark-up
120 400	121 000	0·9950	−0·50	8
63 200	62 000	1·0193	1·93	5
97 400	92 000	1·0586	5·86	2
192 000	200 000	0·9600	−4·00	10
49 000	50 200	0·9760	−2·40	9
73 000	70 400	1·0369	3·69	4
40 200	36 500	1·1013	10·13	1
156 000	155 700	1·0019	0·19	7
102 300	98 200	1·0417	4·17	3
89 800	88 400	1·0158	1·58	6

Column 1   Lowest competitive bid, that is, the winning bid or in the case of contracts that we won, the second lowest bid

Column 2   Our estimate of the cost

Column 3   Is the lowest competitive bid expressed as a proportion of our estimate of the cost, that is it is column 1 divided by column 2

Column 4   Indicates the mark-up needed to win the contract expressed as a proportion of our estimate of the cost; it is calculated by deducting 1·00 from the figure in column 3

Column 5   Gives the rank order of magnitude of mark-up needed to win, that is, the highest figure appearing in column 4 is given the number 1 in column 5, the second highest number in column 4 is given the number 2 in column 5, and so on

It can be seen from table 10.9 that only one of the ten contracts listed would have been won if the mark-up used expressed as a proportion of our cost estimate had been of a value of 10 per cent. Two contracts out of ten would have been won if the mark-up had been 5 per cent, and so on. Thus it is possible to obtain some idea of the relationship between mark-up and success rate and hence carry out the calculation of expected value of profit as shown in table 10.10.

In this way the methods suggested by Gates can be used as in the previous method to calculate the expected value of profit and hence determine the optimum level of mark-up. It is possible, however, to use the results of this calculation in a slightly different way; for example in the case of a particular contract it may be required to increase the company's chance of success in bidding rather than to maximise the expected value of profit. This could occur at a time when a company had a low order book when it might be thought desirable to take on work at low profit simply in order to employ

Table 10.10   Success rate related to mark-up

Mark-up ratio	Probability of success	Expected value of profit (mark-up × probability of success)
−0·0400	1·0	−0·0400
−0·0240	0·9	−0·0216
−0·0050	0·8	−0·0040
0·0019	0·7	0·0013
0·0158	0·6	0·0095
0·0193	0·5	0·0097
0·0369	0·4	0·0148 (Maximum e.v. profit)
0·0417	0·3	0·0125
0·0586	0·2	0·0117
0·1013	0·1	0·0101

company resources usefully. The relationship between mark-up and success rate would enable the company management to determine the appropriate mark-up to achieve the desired success rate. If the company had decided that the minimum acceptable profit level was 4 per cent of turnover then it can be seen from the figures given that the success rate could not be greater than about one in three tenders.

An important factor to consider when calculations are being undertaken to maximise the expected value of profit is that success will be randomly distributed with time and will result in a fluctuating workload. Many companies, however, cannot accept a fluctuating workload and will have to try to smooth it out by varying their bidding strategy. In this situation the model suggested by Gates may be the most useful since it gives a method by which the success rate can be changed by varying the mark-up level.

One of the arguments in favour of Gates' method is that it requires less information, namely only our own cost estimates, our own bid and the lowest competitive bid. The exclusion of all other competitors' bids may be thought to be a deficiency but in the award of contracts it is only the lowest bid that is taken into account and the others are all rejected. Furthermore it might be argued that the lowest bids will all have been carefully calculated whereas relatively high bids may be only rough approximations.

Simulation

Bidding strategy offers an ideal opportunity for the use of a simulation model. This is a mathematical representation of the real-world bidding situation which simulates the action of a number of competing tenderers. For educational purposes bidding simulation is sometimes run in the form of a business game, but it is possible to produce a simulation that represents fairly closely the situation in which a company finds itself. A description of a bidding game will, however, make clear the principles used, as follows.

The simulation model will need to contain basic information on such things as the number of tenderers and their relative advantages in terms of location or ability, limitations on financial and working resources, and details of a series of contracts to be put out to tender. The details of each contract will usually consist of its estimated cost and its scheduled starting and completion dates. The usual procedure is for a number of teams each representing a tenderer to be given details at the start of a number of contracts coming out to tender in a particular month, including information on estimated cost, duration and location. Each team can then submit a bid for each contract and normally the lowest bid is successful and the work is undertaken by the team making that bid. The model will keep track of the teams' cash flow in and out but will not make this information available to the team, which is expected to keep its own accounts. It is normal practice to go through several rounds of this game so that the team may build up a reasonable order book and modify their bidding policy accordingly. Direct negotiation between teams is usually permitted so that any team with excessive work can sub-contract it to another team which is less busy.

While the type of model described above used as a business game is entirely for educational purposes it is possible to observe the competing tenderers in a real situation and represent their action by simulation model. This model may then be used to assess the effect of different bidding policies that the company may wish to adopt. In this way simulation makes available a limited degree of experimentation in situations where live experimentation could be extremely expensive. This is true not only in the bidding situation for contracts but in many other areas of management activity. It has been used widely to simulate aircraft landing at airports, where the arrival rate of aircraft is known to be random. If, for example, it was desired to close one runway at a busy airport for maintenance purposes it would be possible to determine the effect of doing this by use of a simulation model. This would offer just as good a solution as a more rigorous analysis by queuing theory and would be a lot simpler to operate. Other common simulation models are used in stock-control problems where companies wish to test the effect of different levels of stock holding when faced with random orders from customers. Simulation is indeed a very powerful tool in situations that are so complex that fully detailed analysis is almost impossible.

Business Games and Game Theory

These two similar phrases have quite different meanings. A business game is essentially a simulation as described above and presents an opportunity to try to create a competitive situation for the training of managers. Game theory is quite distinct and is a well-developed body of knowledge evolved from a careful analysis of strategy in times of war. It has been extended to other areas of activity such as business strategy but it is really beyond the scope of this book and furthermore has rarely been used in the construction industry.

References

1. H. I. Ansoff, *Corporate Strategy*, Penguin, London, 1970.
2. H. I. Ansoff., ed., *Business Strategy*, Peguin, London, 1969.
3. M. Gates, *Bidding Strategies*, American Society of Civil Engineers, Construction Division, 1967.

Appendix

Present value of £1 discounted at r per cent for n years

Discount rate r per cent

n years from now	6	8	10	12	14	16	18	20	25	30	35	40
1	0·943	0·926	0·909	0·893	0·877	0·862	0·847	0·833	0·800	0·769	0·741	0·714
2	0·890	0·857	0·826	0·797	0·769	0·743	0·718	0·694	0·640	0·592	0·549	0·510
3	0·840	0·794	0·751	0·712	0·674	0·641	0·608	0·579	0·512	0·455	0·406	0·364
4	0·792	0·735	0·683	0·636	0·592	0·552	0·516	0·482	0·410	0·350	0·301	0·260
5	0·747	0·681	0·621	0·567	0·519	0·476	0·437	0·402	0·328	0·269	0·223	0·186
6	0·705	0·630	0·564	0·507	0·456	0·410	0·370	0·335	0·262	0·207	0·165	0·133
7	0·665	0·583	0·513	0·452	0·400	0·354	0·314	0·279	0·210	0·159	0·122	0·095
8	0·627	0·540	0·467	0·404	0·351	0·305	0·260	0·233	0·168	0·123	0·091	0·068
9	0·592	0·500	0·424	0·361	0·308	0·263	0·225	0·194	0·134	0·094	0·067	0·048
10	0·558	0·463	0·386	0·322	0·270	0·227	0·191	0·162	0·107	0·073	0·050	0·035
11	0·527	0·429	0·350	0·287	0·237	0·195	0·162	0·135	0·086	0·056	0·037	0·025
12	0·497	0·397	0·319	0·257	0·208	0·168	0·137	0·112	0·069	0·043	0·027	0·018
13	0·469	0·368	0·290	0·229	0·182	0·145	0·116	0·093	0·055	0·033	0·020	0·013
14	0·442	0·340	0·263	0·205	0·160	0·125	0·099	0·078	0·044	0·025	0·015	0·009
15	0·417	0·315	0·239	0·183	0·140	0·108	0·084	0·065	0·035	0·020	0·011	0·006
16	0·394	0·292	0·218	0·163	0·123	0·093	0·071	0·054	0·028	0·015	0·008	0·005
17	0·371	0·270	0·198	0·146	0·108	0·080	0·060	0·045	0·023	0·012	0·006	0·003
18	0·350	0·250	0·180	0·130	0·095	0·069	0·051	0·038	0·018	0·009	0·005	0·002
19	0·331	0·232	0·164	0·116	0·083	0·060	0·043	0·031	0·014	0·007	0·003	0·002
20	0·312	0·215	0·149	0·104	0·073	0·051	0·037	0·026	0·012	0·005	0·002	0·001
25	0·233	0·146	0·092	0·059	0·038	0·024	0·016	0·010	0·004	0·001		
30	0·174	0·099	0·057	0·033	0·020	0·012	0·007	0·004	0·001			
35	0·130	0·068	0·036	0·019	0·006	0·006	0·003	0·002				
40	0·097	0·046	0·022	0·011	0·005	0·003	0·001	0·001				

Index

Page numbers in bold type give the main reference.